How do you predict the parameters of future solar cycles? What is the role of dynamo theory in the cyclic activity of the Sun and similar stars? And what are the implications of chaos theory for stellar cycles? This book answers these questions and offers a timely review of studies of the cyclic activity of the Sun and other stars.

This authoritative reference shows the importance of reliable predictions of the parameters of future solar cycles, and carefully explains the methods currently used to determine these (with special reference to the maximum Cycle 22). Some of the latest research into solar cycles is clearly presented; this includes helioseismology, observations of the extended activity cycle and the polar field reversals, and contributions from dynamo theory and chaos theory.

For graduate students and researchers, this monograph provides a much-needed synthesis of our understanding of activity cycles in the Sun and other stars.

Cambridge astrophysics series

Editors: R. F. Carswell, D. N. C. Lin *and* J. E. Pringle

Solar and stellar activity cycles

Titles available in this series

SOLAR AND STELLAR ACTIVITY CYCLES

PETER R. WILSON

School of Mathematics and Statistics, University of Sydney

CAMBRIDGE
UNIVERSITY PRESS

PUBLISHED BY THE PRESS SYNDICATE OF THE UNIVERSITY OF CAMBRIDGE
The Pitt Building, Trumpington Street, Cambridge, United Kingdom

CAMBRIDGE UNIVERSITY PRESS
The Edinburgh Building, Cambridge CB2 2RU, UK
40 West 20th Street, New York NY 10011–4211, USA
477 Williamstown Road, Port Melbourne, VIC 3207, Australia
Ruiz de Alarcón 13, 28014 Madrid, Spain
Dock House, The Waterfront, Cape Town 8001, South Africa

http://www.cambridge.org

First published 1994
First paperback edition 2003

A catalogue record for this book is available from the British Library

Library of Congress cataloguing in publication data

Wilson, Peter R.
Solar and stellar activity cycles / Peter R. Wilson.
p. cm. – (Cambridge astrophysics series: 24)
Includes index.
ISBN 0 521 43081 X hardback
1. Solar cycle. 2. Stars. 3. Astrophysics. I. Title.
II. Series.
QB526.C9W55   1994
523.7–dc20   93-29835   CIP

ISBN 0 521 43081 X hardback
ISBN 0 521 54821 7 paperback

Although primarily an observer and experimentalist, Hale brought his acute physical insight to bear on astronomical problems and saw Solar Physics as an integral part of the study of Astrophysics. This volume is dedicated to the elaboration of that proposition.

Contents

Preface

What seems unimaginable in the future becomes inevitable in the past.
Anon.

In February of 1985, an international group of solar astronomers, including both observers and theoreticians, met in Tucson, Arizona, and agreed to plan a series of workshops with the aim of mounting a coordinated study of the new solar cycle, Cycle 22, which was expected to begin in 1986. Meetings were hosted by the California Institute of Technology at the Big Bear Solar Observatory in August of 1986, by Stanford University at Fallen Leaf Lake in May 1987, by the University of Sydney in Sydney, Australia, in January 1989, and by the National Solar Observatory in Sunspot, New Mexico, in October 1991.

This volume does not seek to provide a formal account of the workshop proceedings, which may be found elsewhere. It has, however, been inspired by the intellectual stimulation generated by these meetings and by the many contacts with scientists throughout the world which have followed them.

While making full acknowledgment to the many people whose work and ideas have provided me with excitement and stimulation, I do not wish to imply that this book represents a general consensus. It is not possible to provide a definitive account of the mechanisms underlying cyclic activity at the present time; opinions differ strongly on some aspects, whereas a general bafflement prevails in other areas. It is thus an exciting field, and this volume is intended to set forth a summary of the current state of our understanding of stellar cycles, as interpreted by the author. Where opposing views are strongly held, I have endeavoured to identify and reference them, but the book is frankly biased towards my views and those of my collaborators.

This should not be taken to imply an assertion that these views are right, for I am very conscious that there is much we do not understand. Rather, I believe that it is more useful at this stage to attempt to combine all the phenomena of cyclic activity into a loose structure with a common theme and viewpoint. Future work may then confirm and extend this structure, or it may discard and replace it.

Chapters 1 to 7 and 14 to 15 are intended to be accessible to the interested layperson. Chapters 8 to 13 contain some mathematical or research material which should be accessible to a commencing graduate student in the physical sciences and, together with the earlier material, provide the student with the necessary background and information to undertake, with some guidance, a research program in this area. Those sections related to current or recent research are printed in smaller type and may be skimmed over by the more general reader.

For the expert in the field I hope that a structure will emerge from this attempt to bring together most of the current observational and theoretical aspects of cyclic activity. Although not the final answer, this structure should provide a reference frame for further researches in the field during Cycle 23.

Acknowledgments

Thanks and acknowledgment are due to many people. To Hal Zirin and the staff of the Big Bear Solar Observatory I am grateful for frequent hospitality and for intellectual stimulation. In addition to Zirin, Sara Martin, Bill Marquette (BBSO), Karen Harvey, Douglas Rabin (NSO, Tucson), Dick Altrock (NSO, Sunspot), Rick Bogart, Ron Bracewell, Phil Scherrer, and Todd Hoeksema (Stanford), Elizabeth Ribes (Meudon), and Ann Cannon and Chris Durrant (Sydney) all played major roles in organizing the several workshops, leading the different working groups and contributing significantly to the ideas generated therein. I am also grateful to Tom Holzer and the staff of the High Altitude Observatory for their hospitality during this project.

Discussions with many other people have been of great benefit to me in developing and testing many of the ideas included here. Mark Giampapa and Rich Radick provided generous comments on the chapter on stellar cycles; Frank Hill, Douglas Gough, Phil Goode, Ken Libbrecht and Martin Woodard did likewise on helioseismology; Jack Harvey, Bob Howard and Neil Sheeley (and the members of his group) on the evolution of the large-scale magnetic features; and Nigel Weiss, Gene Parker, and David Galloway on dynamo theory. Many discussions with Peter Fox on almost all aspects of cyclic processes have been particularly valuable. In identifying and thanking these people I do not wish to imply that they necessarily endorse everything contained herein; indeed, Sheeley, Howard and I have disagreed on many issues. I have, however, appreciated their willingness to discuss these matters and, in Sheeley's case, to run several parallel simulations. I also thank Norman Murray for his collaboration on the study of the polar magnetic fields and my secretary, Ms. K Yamamoto, and my research assistant, John Giovannis, for their invaluable help in preparing the manuscript and the figures.

At many points I am indebted to the authors of the many excellent books and review articles on the various topics covered in this study, in particular to Bray and Loughhead (1964), in regard to Chapter 2; to Foukal (1990) and Stix (1989) (Chapter 3); to Priest (1982) (Chapter 4); to Olin Wilson and his collaborators (Wilson *et al.* 1981) and Radick (1992) (Chapter 7); Harvey (1992) (Chapter 8); and to Rosner and Weiss (1992) and Cowling (1981) in regard to Chapter 11 (see references cited in those chapters).

Three people, however, deserve my special acknowledgment and appreciation: Pat McIntosh of the Space Environment Laboratory, NOAA, Boulder, CO; Herschel Snodgrass of Lewis and Clark College, Portland, OR; and Jack Thomas of the University of Rochester, NY. Many of the ideas expressed here have evolved over nearly 30 years of collaboration with McIntosh, and our joint paper of 1985 on 'A new model for flux emergence and the evolution of sunspots and the large-scale fields' provided the genesis for some of the concepts explored in this book. I first met Snodgrass at the Big Bear meeting of the Solar Cycle Workshop in 1986, and we quickly discovered that we held similar views on the torsional oscillations and the possibility of the existence of giant doughnut-shaped convection rolls. This has led to a fruitful collaboration, and I am particularly grateful to Snodgrass for his advice regarding the manuscript during his visit to Sydney in the first half of 1992, and for preparing an excellent summary of Chaos Theory as it relates to cyclic phenomena, which forms the basis of Chapter 13. I am indebted to Thomas for his meticulous job of editing the penultimate draft during his visit to Sydney in 1992–93, for valuable discussions on many theoretical points, and for his generous advice regarding presentation.

Finally I thank my wife, Geraldine Barnes, without whose shining example of single-minded determination to complete her own book, *Counsel and Strategy in Middle English Romance* (D. S. Brewer, Cambridge), particularly while accompanying me on the astrophysical byways of the south-western United States, this book might never have been attempted. Its completion owes much to her ready encouragement, generous assistance with proof-reading, and particular attention to matters of punctuation.

1
Introduction

'Where shall I begin, please your Majesty?' the White Rabbit asked.
'Begin at the beginning,' the King said, very gravely...

Lewis Carrol

1.1 The significance of stellar activity cycles

During the twentieth century, our perception of the fundamental nature of
stars and stellar systems has undergone a revolution almost as profound as
that initiated by Copernicus in relation to the solar system. In the nineteenth
century, a star was regarded as a luminous, spherically symmetric system, for
which the only available energy source appeared to be the energy released by
gravitational contraction. Unfortunately, simple calculations showed that, on
this basis, the Sun's luminous lifetime (the Kelvin–Helmholtz time) was far
too short to accommodate the age of geological structures, the development
of life, and the evolution of species.

The discovery that nuclear energy could provide the source necessary
to prolong the luminous lifetimes of stars by several orders of magnitude
was the first significant development in our understanding and provided
the background structure for the picture of a star that emerged in the
first half of this century: i.e. that of an *equilibrium system*, in which the
internal generation of nuclear energy remained in long-term balance with
the radiation emitted at the surface. In this system, it was assumed that
hydrostatic pressure balance applied and that the outward temperature
gradient was monotonically negative, in conformity with the well-understood
principles of thermodynamics.

This comfortable picture did not, however, survive into the second half
of the twentieth century. Gradually, astrophysicists were forced to accept
that a star is a *non-equilibrium system*, from which energy is lost in the

1

form of both matter and radiation. The expected monotonically negative outward temperature gradient was replaced by a steep temperature rise through a chromosphere and corona. These characteristics of the outer solar atmosphere were at first thought to be peculiar to the Sun but are now recognized as properties of many, if not most, late-type stars.

Again, while *solar activity*, in the form of sunspots, flares, and related phenomena has long been known, the discovery that such activity occurs on most stars, and that it frequently appears in more violent forms on stars other than the Sun, prompted a further reassessment of the nature of stars. When solar studies revealed that activity is a manifestation of the interaction between plasma motions and magnetic fields, the widespread occurrence of the phenomena of activity provided evidence for the almost universal presence, and importance, of *stellar magnetic fields*.

This discovery represents the next major development in our understanding, for, although the energy associated with the magnetic fields is generally far less than that stored in the convective or rotational motions of the stellar plasma, a variety of mechanisms are now known whereby these fields can exert a disproportionate influence on the motions and on the radiation emitted from the surface. Thus, whereas formerly it was believed that the structure of an equilibrium star could be determined, in principle, once its mass, chemical composition and age were given, it is now accepted that real stars of similar masses and compositions may exhibit remarkably different atmospheric (and probably internal) structures and activity phenomena because of the varied effects of the magnetic fields which they possess.

The realization that magnetic fields and their interaction with moving plasmas might hold the key to phenomena as diverse as small ephemeral regions and spicules on the Sun and the structure of galaxies has, of course, concentrated considerable attention on this interaction, the many facets of which continue to challenge our understanding. A dramatic example, and one of considerable importance to this study, is the sunspot. Although sunspots can be clearly resolved on the solar surface and have long been the subject of intense investigation, their true nature continues to elude us. We have progressed from regarding them as blemishes on an otherwise perfect sphere, to holes in the solar atmosphere through which the (then believed to be) cooler solar interior could be viewed, to solar surface tornados, and, finally, to the modern view in which they are recognized to be regions where powerful concentrations of magnetic field emerge through the surface. Nevertheless, as E. N. Parker (1975) succinctly expressed it, 'sunspots are too unstable to form and, if once formed, should immediately break apart.

But observations show the contrary, indicating that there is much we do not understand'.

An intriguing aspect of sunspots is the 11-year variation of their annually averaged number count, which is known as the *sunspot cycle*. Although sunspot number counts had been carefully monitored at certain observatories since the early 1700s, it was not until 1843 that the German apothecary and amateur astronomer, Heinrich Schwabe, discovered that their comings and goings exhibited a well-defined period of ~ 10 years. More recently, it has been recognized that the period is closer to 11 years and that the frequency of occurrence of the related activity phenomena also fluctuates with a similar period. Today, the term *solar activity cycle* is now used to embrace the cyclic variation of all such phenomena.

In Schwabe's time, the sunspot cycle was regarded as a curiosity. It was not until the present century that it was recognized as one facet of a more general type of stellar phenomenon. While certain classes of variable stars (e.g. the Cepheids, and the RR-Lyrae stars) were recognized in the nineteenth century, it was generally believed, and not only among scientists, that the majority of stars were unchanging in their properties, at least on human timescales. ('Bright star! would I were steadfast as thou art', wrote the poet, John Keats). The recognition that variability is a universal property of stars, that it may appear in many forms, and that the sunspot cycle is simply a particular example was yet another development in our understanding during the present century.

While the Cepheids and the RR-Lyrae stars vary in their total luminosity, sometimes by an order of magnitude, the sunspot cycle is principally an activity variation. It is now known, however, that luminosity variations, albeit of considerably smaller amplitudes, also accompany stellar activity variations, and that they are sometimes in phase and sometimes out of phase with the activity variations.

Thus, today, we recognize that the sunspot cycle is simply an example of *cyclically varying stellar activity*, the subject of this volume.

1.2 The solar–stellar connection

An essential ingredient of this study is an understanding of the relationship between solar and stellar activity cycles, which is an important component of a more general methodology known as the *solar–stellar connection*.

Most of our information comes, of course, from the Sun, for which the full disk can be resolved, and our growing knowledge of solar activity provides direction for our study of activity in other stars. Stellar investigations, however, are essential to an understanding of cyclic activity phenomena, for only in this way can we investigate the dependence of stellar activity on parameters such as age, size, surface temperature, rotation rate, and convection zone structure, which, for our studies of the Sun, are necessarily fixed. Since *cyclic activity* is a property of some, but by no means all, active stars, a knowledge of the properties of those active stars which behave in this way, and of those which do not, should be of importance to an understanding of the cyclic phenomenon.

1.3 The solar cycle and the terrestrial environment

Quite apart from contributing to our understanding of the mechanics of stellar activity, the cyclic behaviour of the Sun is of special importance for our terrestrial environment. The effect on radio communications of particle streams incident on the ionosphere is well known, as is the apparent relationship of these streams to solar flares, whose number and intensity vary with the cycle. On 10 March 1989, an X-flare gave rise to a geomagnetic storm two days later which blacked out the Hydro-Quebec power system. Transformers failed, hundreds of relay and protective systems malfunctioned, and voltage and power fluctuations were widespread across North America.

Dosimeters registered alerts on high-flying Concordes during the October 1989 proton event. On long intercontinental flights in the auroral zones a 'chest X-ray' dosage of radiation can occur. So seriously does NASA regard these interactions with the geosphere in the planning of the space program, that it is an active sponsor of research aimed at more accurate predictions of the parameters (amplitude, period, etc.) of the cycle. Nevertheless, the only flare-research spacecraft programs actively underway are the Japanese *Solar-A* and the Russian *Coronas*.

Of considerably greater long-term importance to our environment, however, is the variation of the total radiative output of the Sun during the cycle. One might intuitively expect that, when the Sun is covered with a larger number of cooler spots, its radiative output should decrease. Satellite radiometer measurements since 1980 have, however, shown that the output actually declined until 1986, coinciding with sunspot minimum, and that, since then, it has undergone a steady increase (at least until 1990). Although the amplitude of the increase is less than one per cent, corresponding to a temperature change of $\sim 0.2\,\mathrm{K}$, it is likely that, because of the non-linear nature of the interactions, the climatic effects could be considerably greater, particularly if changes in the solar output give rise to non-uniform temperature variations over the geosphere.

For example, it has been shown (J. Lean, private communication) that the temperature differential between land and ocean on the eastern seaboard of the United States tracks the sunspot cycle very closely. Although the actual differential is small, it may give rise to alternating wind patterns which may cause significant changes in the local climate. Indeed, correlations between terrestrial weather patterns and the sunspot cycle have been claimed. Xanthakis (see Giovanelli 1984) finds that the mean precipitation in the northern hemisphere latitude range N70° − 80° has varied in phase with the sunspot cycle from 1880 to 1960, while that in the range N60° − 70° has varied out of phase. In the southern hemisphere Bowen (see Giovanelli 1984) finds that, over the same period, the rainfall at Hobart (Tasmania, latitude S43°) is greatest at sunspot maximum, while that at Cairns (Queensland, latitude S17°) is least.

Since the time-span of these records covers only a few cycles, these results should be regarded with some caution. It is interesting, however, that, between 1640 and 1705, remarkably few sunspots were observed, and the Earth's climate suffered a significant cooling during the same period, now referred to as the 'Little Ice Age'. If there is a physical connection between these two excursions from the norm (and it is now generally accepted that there is), the need to be able to make reliable predictions about likely future climatic variations resulting from solar variability reinforces the need to understand the nature of stellar activity cycles.

There are many other terrestrial phenomena for which links with the sunspot cycle have been claimed, sometimes on rather dubious scientific grounds. In 1937, Harlan T. Stetson, then a research associate at MIT, published a volume entitled *Sunspots and their Effects*, in which links between the sunspot cycle and variations in (i) human behaviour, (ii) agriculture, (iii) radio communication, (iv) business and the stock market, (v) the weather, (vi) geomagnetic phenomena, and (vii) the performance of carrier pigeons were claimed. For some of these, such as radio communication and geomagnetic phenomena, the evidence is well documented and the mechanisms at least partly understood. Since there is some evidence that weather patterns vary with the cycle, it is possible to understand how agricultural output may also be related not only to normal sunspot fluctuations but also to gross excursions, such as occurred during the Little Ice Age.

The possible links with human behaviour, business, and the stockmarket are far more speculative. Stetson shows several figures, some of which are reproduced in Figure 1.1, in which sunspot number fluctuations during the 1920s and 1930s are compared with business activity, automobile production, and building contracts of that period. At first glance, these rather picturesque

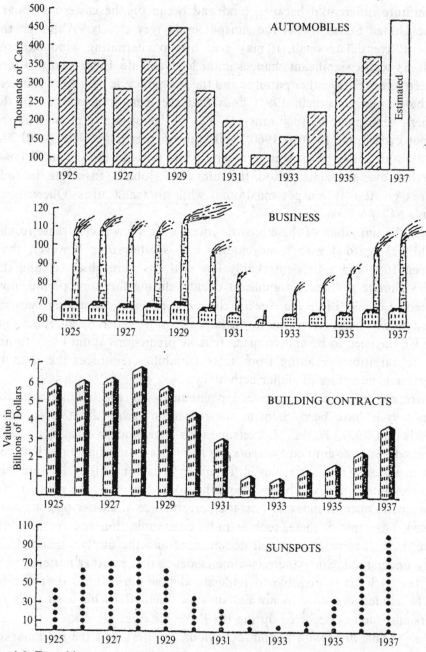

Fig. 1.1 Four histograms, taken from Stetson (1937), compare automobile production in the US, business activity (in unspecified units), and the value of building contracts written in the period 1925–37 with the sunspot number during that period.

histograms appear to show a remarkable correlation, but the coincidence of the sunspot minimum of 1933 with the trough of the Great Depression may well be just that: a coincidence. Indeed, the fact that the current world recession (1989–91) following the stock market crash of 1987 coincided with the *maximum* of Cycle 22 does *not* support the likelihood of such a connection.

An even more bizarre claim of a correlation between sunspot number and human mortality was made a few years earlier by a German physician, Bernard Dull, and his wife, Traute (1934). From a study of all the deaths in the city of Copenhagen between 1928 and 1934, they claimed to have detected a 27-day periodicity by averaging over 68 successive 27-day periods. They also noted a parallel with a 27-day fluctuation in sunspot numbers arising from solar rotation and a concentration of sunspots in what they call 'the M-zone', and claimed that a relation was established. The statistics of the study were less than impressive, however, the amplitude of the mortality fluctuations being only a few per cent, and the standard analytical techniques, such as power spectral analysis (see Chapter 5), unreported. As we shall see, the phase of the 27-day periodicities in solar phenomena is subject to change due to the changing distribution of active regions in the solar surface. No reference was made to cyclic changes in the Dulls' study, although the period in question included the minimum of Cycle 17. Modern medical statisticians find no correlation between human mortality and the sunspot number, and, as a result of these and similar claims, sunspot correlation studies became somewhat discredited.

It would nevertheless be unwise to ridicule the possibility of connections between the sunspot cycle and various fields of human activity without investigation. The Sun is such an integral part of our environment that its effects are frequently taken for granted, and we should not reject *a priori* the possibility of some forms of interaction. The preferred approach is to assess the reliability of the data-gathering process and the statistical significance of the correlations before attempting to underpin any such correlations with a physical theory (see further Chapter 14).

1.4 Conclusion

The aim of this volume is to explore the phenomena of solar and stellar activity and activity cycles and to attempt to understand the physical processes that govern them. By doing so, it is hoped to achieve a greater understanding of the relationship between solar activity and its cyclic variation and our terrestrial environment.

References

Dull, Bernard, and Dull, Traute: 1934, in *Virchow's Archiv für Pathologische Anatomie.*

Giovanelli, R. G.: 1984, *Secrets of the Sun*, Cambridge University Press, Cambridge, UK, 109.

Parker, E. N.: 1975, *Solar Phys.* **40**, 291.

Stetson, H. T.: 1937, *Sunspots and their Effects*, Wittlesey House, McGraw-Hill Book Co., New York.

2

Historical survey

Those who will not study history are condemned to repeat it
Karl Marx

History is bunk
Henry Ford

2.1 The discovery of sunspots

Although naked-eye observations of sunspots have been recorded sporadically since the first Chinese observations several centuries before the birth of Christ, the year 1611, when sunspots were observed for the first time through the telescope, marks the beginning of the science of astrophysics. Four men share the honour of this discovery: Johann Goldsmid in Holland (1587–1616), Galileo Galilei in Italy (1564–1642), Christopher Scheiner in Germany (1575–1650), and Thomas Harriot in England (1560–1621). It is uncertain which of this international quartet made the first observations, but priority of publication belongs to Goldsmid, or Fabricius, as he is known by his Latinized name. Although his equipment was probably inferior to that of Galileo or of Scheiner, Fabricius made observations of sunspots and used them to infer that the Sun must rotate but did not carry this work beyond these initial observations.

When Scheiner, a Jesuit priest teaching mathematics at the University of Ingolstadt, first observed the spots, he suspected some defect in his telescope. He soon became convinced of their actual existence but failed to persuade his ecclesiastical superiors, who refused to allow him to publish his discovery. This indignity was later shared by the French astronomer, Messier, who in 1780 was similarly prevented from announcing his observation of the return of Halley's comet in that year. Regrettably, such instances of scientific censorship are not uncommon and, in Scheiner's case, played a major role

9

in the controversy that led to the denouncement of Galileo to the Italian inquisition.

Unable to publish, Scheiner, instead, announced his discovery in three anonymous letters to Mark Wesler, a friend of Galileo. Wesler, in turn, showed the letters to Galileo, and Galileo quickly responded in three famous letters (*The Sunspot Letters*, written in 1612), claiming priority of discovery and giving an account of his own researches. Whether Galileo had actually observed sunspots before seeing Scheiner's letters is not known, but Scheiner was infuriated with Galileo's belated claim of priority. This incident and the controversy that grew out of it made Scheiner a bitter enemy of Galileo and certainly contributed to the hostility with which Galileo was regarded by the Jesuits. The famous quarrel and its various implications are still a topic of interest to historians of science (see, for example, Drake 1957, 1990).

Galileo pursued his researches into sunspots for only two years but did so with the characteristic enthusiasm and skill through which he laid the foundation for the scientific method. He crushingly disposed of Scheiner's suggestion that the spots, or blemishes, might really be small planets, by showing that this hypothesis was quite incompatible with their observed changes in size and shape, further fuelling Scheiner's wrath. Galileo also inferred that the Sun must revolve about a fixed axis with a period of about one lunar month and noticed that spots within a single group move relative to one another.

Undaunted, Scheiner vigorously continued his sunspot observations for nearly two decades and finally published his work in 1630 (*Rosa Ursina sive Sol*). He noted, as had Galileo, that the spots occurred within zones of low latitude at either side of the equator but never near the poles. Scheiner also noted that spots at higher latitudes rotated more slowly and that the axis of the rotation was tilted with respect to the ecliptic. He also made detailed drawings in which the umbra and penumbra were clearly distinguished.

The English member of this quartet, Thomas Harriot, did not become embroiled in these controversies. Once mathematical tutor to Sir Walter Raleigh, Harriot accompanied Sir Richard Grenville to Virginia as a surveyor and, on his return, was rewarded by being introduced to the Earl of Northumberland, under whose patronage he was able to devote the rest of his life to the pursuit of scientific endeavours. He apparently had access to one of the first telescopes and reported his observations of sunspots during the year 1611, but any further achievements on Harriot's part are not recorded.

2.2 The Maunder Minimum

After the initial flurry of interest and the publication of Scheiner's major work, the study of sunspots went into an apparent decline. Only recently has the work of Eddy (1977, see also Pepin *et al.* 1980) shown that there was a genuine reduction in the number of sunspots during the period 1640–1705, which is now known as the 'Maunder Minimum'. The Maunder Minimum will be discussed in detail in Chapter 5. Here we merely note that it corresponded to a climatic excursion in Europe, known as the 'Little Ice Age'.

2.3 The periodicity law

Although sunspots began to reappear with greater frequency after 1705, it was not for nearly another 150 years that the next significant discovery was made. Heinrich Schwabe (1789–1875) was a German apothecary whose hobby was astronomy. He was born in Dessau, studied at the University of Berlin, and then returned to Dessau to commence his business. In 1826, he purchased a small telescope and began solar observations, with the object of searching for a planet inside the orbit of Mercury. Like Messier, who faithfully recorded the coordinates of small patches of nebulosity in the hope of discovering a successor to Halley's comet, only to provide science with its first catalogue of galaxies, Schwabe recorded the occurrence of sunspots for 43 years in order to assist his search. He failed to locate the hypothetical planet but noted, from his records, a probable 10-year periodicity in the occurrence of sunspots. His first announcement, in 1843, attracted little attention, but he continued his observations and, in his famous treatise, *Kosmos*, published in 1851, Humboldt included Schwabe's table, which clearly demonstrated the 11-year periodicity of annually averaged sunspot numbers.

It is curious that it took so long for this periodicity, so obvious in the plot of the Zurich sunspot numbers (see § 2.5) shown in Figure 2.1, to be noted. Present studies of sunspot records kept stored at the Observatoire de Paris indicate that, even during the Maunder Minimum, there were some spots on the Sun, enough in fact to see, with hindsight, that there was a cyclicity in their comings and goings. One suspects that, while the records were carefully kept, they were never examined or plotted as in Figure 2.1. It took an amateur with a specific objective in mind, who scrutinized his observations with great care, to make an entirely different discovery. Not

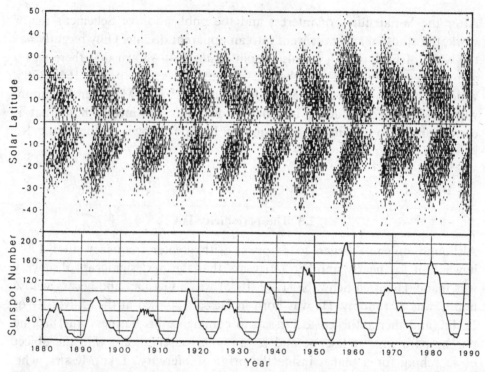

Fig. 2.1 The upper panel shows a plot of the latitude of emergence of individual sunspots against time (in years) since 1880, otherwise known as the *butterfly diagram*. The lower panel shows a plot of the annually averaged Zurich sunspot number R_Z against time.

for the first time in science did a systematic search, guided by one objective, yield an entirely different, and probably more valuable, result.

2.4 The Carrington rotation and the solar flare

News of Schwabe's discovery greatly interested an Englishman, Richard Carrington, son of a wealthy brewer. His father had intended a career in the Church for him, but, while at Trinity College, Cambridge, Carrington's enthusiasm was captured by astronomy. To gain experience, he worked for three years as an assistant at Durham Observatory and then built his own private observatory at Redhill in Surrey. His telescope was an equatorially mounted 4.5-inch refractor which projected an image of the Sun onto a screen. During the period 1853–1861, he made a vast number of sunspot observations and discovered that the average latitude of sunspots decreased from the beginning to the end of the cycle. From his observations of sunspot rotation rates at different latitudes, Carrington deduced that the Sun rotated differentially, a point at the equator rotating more rapidly than one at higher latitudes.

Carrington defined an arbitrary reference point on latitude 16° as longitude zero, and a rotation completed by this point became known as a 'Carrington rotation' (CR). The 'Carrington longitude' of any point on the Sun is the longitude of the intersection of the meridian through that point with the parallel of latitude at 16°, relative to the reference point.

Another important contribution was Carrington's discovery of the *solar flare*. While engaged in sketching sunspot projections in September 1859, he and a companion astronomer, R. Hodgson, saw two crescent-shaped patches break out, brighten, move a distance about twice their length, then fade away as two dots, all within a period of five minutes. This, it turned out, was the first observation of a major flare, and the first indication that sunspots are associated with *solar activity*. Carrington subsequently reported to the Royal Astronomical Society that 'the magnetic instruments at Kew were simultaneously disturbed to a great extent...4 hours after midnight there commenced a great magnetic storm'. However, he added that: 'while contemporary occurrence may be worth noting, I would not have it supposed that I lean towards hastily connecting them. "One swallow does not a summer make"'.

Carrington's caution was commendable, as was his ability to see the possibilities of the interconnections between apparently unrelated events. In this case he had observed an extremely rare phenomenon, a *white light* flare, but confirmation of the flare event and its connection with terrestrial magnetic disturbances had to await the invention of the spectrograph nearly a century later (see § 3.8).

Regrettably for science, the death of his father in 1858 forced Carrington to abandon his observations in favour of the management of the brewery which bears his name. Which career more greatly benefited humankind is open to question, but it is gratifying that Carrington's name is remembered by the reference standard of solar longitude as well as by a brewery.

2.5 The Zurich sunspot number

Confirmation of Schwabe's discovery came when Rudolph Wolf (1816–1893) of Berne searched through all available records and derived a more accurate estimate of 11 years for the average duration of the sunspot cycle. In 1848,

Wolf introduced the relative (Zurich) sunspot number, R_Z, defined by

$$R_Z = K(10g + f), \qquad\qquad (2.1)$$

as a measure of sunspot activity. Here g is the number of sunspot groups, f is the number of individual spots, and K is a personal reduction coefficient, of order unity, allocated by Wolf to represent the reliability of the observer and his equipment. (For his own observations, Wolf took $K = 1$.) Appointed director of the Zurich Observatory in 1855, Wolf instituted a program for the daily determination of R_Z and to this day Zurich remains the world centre of sunspot number data. In 1882, Wolf's successors changed the counting method and, from simultaneous observations using both methods, derived a new value of K, namely 0.6, which reduced the new observations to the old scale.

In order to fill in gaps in his records due to inclement weather or instrumental failure, Wolf initiated one of the earliest international observing networks. Today, more than 30 observatories cooperate by sending their sunspot statistics to Zurich, where they are reduced to the Zurich scale by the determination of the coefficient K for each observatory.

Wolf's definition of R_Z is quite arbitrary, and considerable judgement must be exercised in its measurement; even under identical instrumental and 'seeing' conditions, different observers may obtain different values for f and g. (The term 'seeing', as used by astronomers, is a technical term describing the properties of the atmosphere in the neighbourhood of an observatory as they relate to the quality of the observations obtained. See Bray and Loughhead 1964, p. 18.) An alternative to R_Z as a measure of the sunspot population is the total area of spots visible on the hemisphere, and this has been measured photographically at the Royal Greenwich Observatory since 1874. Other measures have been proposed from time to time, and any attempt to differentiate among these prompts the question as to what is the fundamental cyclically varying physical quantity to be measured, an issue further discussed in Chapter 5. The Wolf sunspot number R_Z has, however, the considerable advantage over all other measures in that it provides the longest database using observations obtained by a coordinated program.

The daily value of R_Z does not, of course, represent the level of activity on the Sun on any one day. It measures, at best, only the spots present on the visible hemisphere and, because of the inhomogeneous distribution of spots over the surface, will vary from day to day as the Sun rotates. Better representations of activity levels are obtained by averaging R_Z over a Carrington rotation or over a year. A plot of the annually-averaged Zurich sunspot numbers since 1880 is shown in Figure 2.1 (see also Chapter 5, Figure 5.1).

2.6 Latitude drift and the butterfly diagram

The latitude drift discovered by Carrington was investigated in more detail by the German astronomer, Spörer, and is occasionally known as 'Spörer's Law'. Spörer began to observe sunspots in 1860 in order to determine the law of solar rotation. He built a small solar observatory at Anclam, in Pomerania, and continued his sunspot observations there until the end of 1873, when he transferred his work to the Potsdam Astrophysical Observatory.

A more dramatic representation of the latitude drift data was given by Maunder, in 1922, when he published them in the form of his *butterfly diagram*, a presentation which convinced scientists that this drift contained an essential clue to the nature of the cycle. A current version of the diagram for the cycles since 1880 is also shown in Figure 2.1.

Distinct differences between individual cycles can be seen in both diagrams. Another important result revealed by the butterfly diagram is that, although the average time between successive minima is ~ 11 years, the average length of each cycle, as determined by the projection of its 'wing' on the time axis, is ~ 12–13 years; that is, successive cycles overlap by ~ 1–2 years. The significance of this result was scarcely appreciated at the time but is further discussed in Chapter 8.

2.7 The Wilson effect

Thus far we have discussed only the numbers, locations, and drifts of sunspots on the face of the Sun. The history of investigations into the nature of sunspots is, however, equally interesting. From the time that Galileo disposed of the idea that they were inferior planets, little progress was made into the physical nature of a sunspot until a discovery by Alexander Wilson, Professor of Astronomy at the University of Glasgow. In 1769, nearly three-quarters of a century before Schwabe's discovery of periodicity, Wilson was attracted by a very large spot nearing the west limb and noted that the penumbra on the side further from the limb gradually contracted and finally disappeared. When the spot reappeared at the east limb some two weeks later, the same behaviour was displayed by the penumbra on the opposite site of the spot, again the one further from the limb.

This phenomenon was subsequently observed in other spots and became known as the 'Wilson Effect'. The explanation suggested by Wilson was that sunspots represented saucer-shaped depressions in the solar surface, formed by the partial removal of the luminous material which, he believed, covered the dark interior of the Sun. While this explanation may appear rather naive today (it was also suggested by Sir William Herschel), it ranks as the first *physical* investigation of the properties of individual sunspots, in contrast to the essentially numerical studies of the occurrence and distribution of spots, which were the main themes of both his predecessors and his successors for upwards of a century. Further, Wilson's hypothesis contained the essentially valid inference that a sunspot may be regarded as a depression in the solar surface, a phenomenon now referred to as the *Wilson depression*.

More recent, but since discarded, theories of the physical nature of sunspots are due to Russell and Rosseland (see e.g. Cowling 1953), who suggested that they arise from the effect of adiabatic cooling on a volume of gas moving upwards through a stably stratified region, and the vortex theory, put forward by Bjerknes (see Cowling 1953), in which the spot is pictured as the seat of a vortex motion around a vertical axis. According to this theory, the vortex motion gives rise to centrifugal pumping, and the resulting adiabatic cooling produces the visible sunspot. The correct explanation of the darkness of sunspots came, however, from a different quarter (see § 3.4).

2.8 Hale and the Mount Wilson era

Probably the most famous name in the history of the investigation of sunspots is George Ellery Hale (1868–1938, see Figure 2.2), designer and first director of Yerkes Observatory, and subsequently of the Mount Wilson Observatory, founder of the California Institute of Technology, and designer of the 200-inch telescope at Mount Palomar. For nearly five decades, Hale was the guiding figure in the design and construction of the world's largest telescopes. He was also extraordinarily active in research and publication, being the founder of two of the leading journals in astronomy and astrophysics, the *Astronomical Journal* and the *Astrophysical Journal*, and a co-founder of the *American Astronomical Society*. During the first three decades of the twentieth century, which saw the great revolutions in physics brought about by scientists like Curie, Einstein, Bohr, and de Broglie, the work of Hale and his co-workers at Mount Wilson was a revolution in its own right, since it changed completely not only our understanding of the nature of sunspots, but also the science of astrophysics as a whole.

Hale invented the solar spectroheliograph in 1890 and, aided by F. Ellerman, constructed one at a small private observatory at Kenwood, Chicago, funded by his father. With this instrument he obtained over 3000 pho-

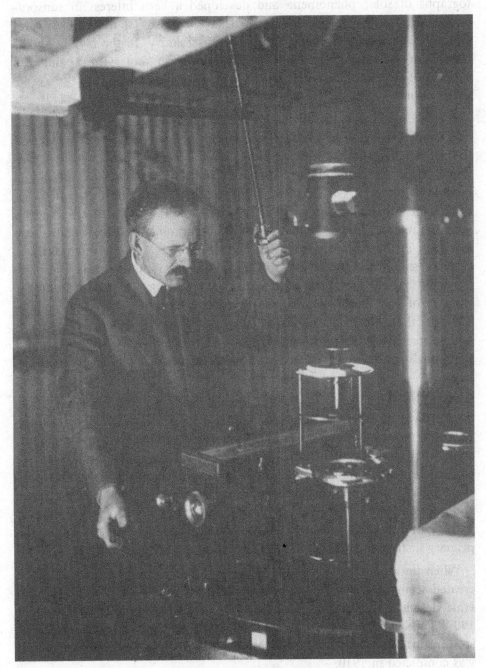

Fig. 2.2 George Ellery Hale.

tographs of solar phenomena and developed a keen interest in sunspots and their spectra. He became an associate professor at the University of Chicago, which took over the Kenwood Observatory. In the mid 1890s the observatory was moved to Williams Bay, Wisconsin, and, with the financial support of streetcar magnate Charles Yerkes, was expanded into the Yerkes Observatory.

It was a lifelong characteristic of Hale that he never stopped planning and developing new systems; as soon as he had completed one project, he was off to the next. He spent the winter of 1903 in Southern California, where he was impressed by the possibilities of Mount Wilson, a peak in the San Gabriel mountains north of Los Angeles, as the site for a mountain observatory for both solar and stellar observations. At that time the lights of Los Angeles seemed far away, the combination of the automobile and the atmospheric inversion in the Los Angeles basin had not yet created the infamous Los Angeles 'smog', and the seeing from this mile-high mountain was unusually good. With great perseverance, he persuaded Andrew Carnegie to fund the transfer of the 'Snow' solar telescope from Yerkes to California, and, in 1904–05, the instrument was carried in sections up the mountain by mules and re-erected at the south-eastern edge of the peak. A spectrograph and a spectroheliograph were installed, and the great period of solar observations began.

Hale's research was concentrated in the field of solar physics. While deeply involved in pioneering work in this field at the start of this century, he continued to develop Mount Wilson Observatory, both as a site for solar work and for night-time astronomy. The funding for most of the construction was supplied by Carnegie, and, until recently, the observatory was run and supported by the Carnegie Institution of Washington. Under Hale's direction, two powerful night-time telescopes were built: the 60-inch in 1907, and the 100-inch in 1917. The 60-inch telescope is today one of the primary instruments used in studies of stellar activity (see Chapter 7).

When he found that ground-level temperature variations and vibrations created seeing problems for the horizontal spectrograph and the coelostat, Hale designed the 60-foot tower telescope, which became operational in 1907. This instrument did not, however, have the desired resolving power, so it in turn served as the prototype for the 150-foot tower telescope, which was completed in 1910.

Hale retired as the director of Mount Wilson in 1923, but his influence and enthusiasm persisted for many years, and the period from 1905 to 1930 became known as the 'Mount Wilson era' in solar physics. Even in his retirement, Hale constructed an observatory in suburban Pasadena and continued to observe and to write (for further reading see Wright *et al.* 1972).

Although primarily an observer and experimentalist, Hale brought his acute physical insight to bear on astronomical problems and saw solar physics as an integral part of the study of astrophysics.

2.9 The magnetic fields of sunspots

The first result from Mount Wilson confirmed what was already suspected. From a comparison of sunspot spectra obtained by the Snow telescope with the spectra of electric arcs produced in the laboratory, Hale showed that sunspots are cooler (~ 4000 K) than the surrounding plasma (~ 6000 K).

The second result was entirely unexpected. Hale had developed the spectroheliograph with the aim of obtaining two-dimensional photographs of the solar surface at a given wavelength. By tuning the spectroheliograph to the wavelength of strong spectral lines originating in the chromosphere (see § 3.6), he could obtain photographs of the Sun at different levels within its atmosphere. Using specially sensitized plates, he obtained the first spectroheliograms in Hα, the strong hydrogen line which originates in the chromosphere, that more tenuous part of the atmosphere where the temperature is increasing outwards.

On some of his photographs of sunspots and their surroundings (see e.g. Figure 2.3, from Hale 1926), Hale found evidence of a vortex structure, which he called 'hydrogen vortices or tornadoes' and, using his considerable physical insight, reasoned as follows:

A magnetic field is produced in the laboratory by an electric current flowing through a coil of wire. If, in harmony with modern physics, free electrons could be assumed to be present in the hot gases of the Sun, a whirling mass of them in a sun-spot vortex should set up a magnetic field. If sufficiently intense, such a field should split the lines in the spectrum of the spot vapours into two or more components, exactly as the same lines are split in the laboratory when these vapours are observed between the poles of a powerful magnet (the Zeeman effect). The components are polarized in a distinctive way, so that one or the other can be cut off by a Nicol prism and quarter-wave plate. Thus the existence of magnetic fields in sun-spots should be unmistakenly recognized.

The observations were made with the recently constructed 60-foot solar tower on 25 June 1908, and the line-splittings duly noted. To his considerable surprise, Hale found that the separation of the components indicated not fields of a few gauss, as he expected, but of the order of several kilogauss. Although his speculation about the source of the field had been perfectly

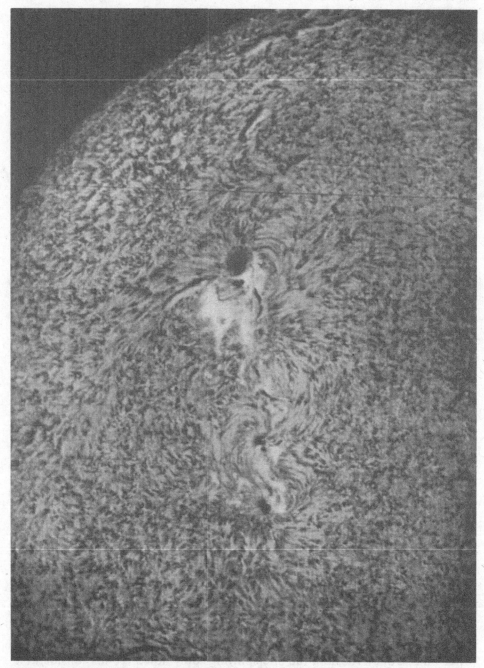

Fig. 2.3 An Hα spectroheliogram showing the structures which Hale described as 'hydrogen vortices or tornadoes in the solar atmosphere above sunspots'. (From Hale 1926.)

logical, it could not account for kilogauss fields. In a footnote to his 1926 review Hale commented: 'Although the magnetic fields are certainly present, the problem is less simple now than it seemed in 1908, because of the difficulty of accounting for the strong electric currents necessary to produce such intense fields.' Indeed, subsequent observations showed that the vortex-like pattern in the Hα spectroheliograms, on which Hale had based his speculation, was the exception rather than the rule. He immediately understood that the explanation of the kilogauss fields which he had discovered must lie elsewhere, and thus solar physics, as distinct from solar astronomy, began. Again, a search guided by one objective yielded an unexpected, and more significant, result.

2.10 The magnetic cycle

Continuing his investigations with this new telescope, Hale noticed, as had Wolf and others before him, that sunspots tend to occur in groups, containing generally two or three larger spots and a number of smaller ones. By convention, the Sun is said to rotate from east to west, and, although the groups exhibit a great diversity of structure, the largest spot of a group tends to be found on the western side and the next largest near the eastern boundary of the group. These became known as the 'leader' and 'follower' spots. In an investigation of the field polarities associated with the spots within such groups, Hale discovered the fascinating sequence of patterns associated with the sunspot cycle:

(i) The leader spots in each hemisphere are generally all of one polarity, while the follower spots are of the opposite polarity.

(ii) If the leaders and followers are regarded as magnetic bipoles, the orientation of these bipoles is opposite in opposite hemispheres.

(iii) The magnetic axes of these leader-follower bipoles are usually inclined slightly towards the equator, the leader spot being the closer.

(iv) Towards the end of a cycle, some spot groups appear at high latitudes with reversed polarity orientation, while those with the normal polarity for the old cycle occur close to the equator.

(v) Following the minimum between the cycles, the bipole polarities all have reversed sign.

In this set of observations Hale thus connected solar minimum with the polarity reversal of the sunspot magnetic fields and identified the high-latitude spots around solar minimum as belonging to the *next* cycle. This confirmed the evidence of the butterfly diagram, that successive cycles overlap but, more significantly, assigned a qualitative difference (the bipole orientation) to the spots of the new cycle. As we shall see in Chapter 8, the overlap of the cycles means that the cycle cannot be characterized by a single parameter, a result which has recently been found to be of considerable significance.

At the time (1914) of these discoveries, Hale suspected that the bipole orientations would reverse sign again at solar maximum, but in this he was disappointed. It was not until the next solar minimum that the bipole re-reversal occurred, restoring the magnetic configurations to those which obtained at the start of the previous but one cycle. In 1924, Hale and his colleague Nicholson (see Hale and Nicholson 1938) were therefore able to announce that the 11-year sunspot cycle was, in fact, a part of a 22-year magnetic cycle.

2.11 The polar fields

Turning his attention to the polar regions of the Sun, in which there were no sunspots, Hale found them to contain magnetic fields with strengths of ~ 50 gauss, but he did not associate these with sunspot fields or with the magnetic cycle. These magnitudes are ~ 10 times the magnitude of the polar fields measured today, which makes one suspect a calibration error. Such an error would not be surprising, given the crudity of the instruments of the time and the weakness of the magnetic fields measured.

As Hale and his co-workers and successors continued to monitor the polar fields, they noted that the strengths measured steadily decreased, but they did not connect this with changes in the sunspot fields or with the cycle. It was not until the 1950s that the next significant discovery was made. Harold Babcock had been brought to the observatory by Hale as a designer of the machinery for the ruling of spectral gratings; when he and his son, Horace, invented the magnetograph, accurate measurements of the fields at the poles and over the entire surface of the Sun became possible.

In 1957, Horace Babcock observed that the north polar fields were of positive polarity, as were the fields of the leader spots of that hemisphere, while, in the south, the polar polarities, as well as those of the leader spots, were negative. As the cycle progressed, however, the polar fields weakened and, by 1960, the polarity of the north polar field had reversed. In the south, the decline of the polar field was slower, so that, for a period of ~ 2 years, the polarity of both northern and southern fields was negative.

It should be emphasized here that the polar fields were weak, a few gauss, even at sunspot minimum, and the 'reversals' not well defined. At times, the polar regions were covered with weak patches of fields of either polarity, and the 'net' polarity of the polar regions was uncertain. Gradually, however, as Cycle 19 declined, the polarity reversals at each pole became more clearly defined, such that, by 1966, the polar regions exhibited fields of clearly negative polarity in the north and of positive polarity in the south. When the active regions of the new cycle, Cycle 20, emerged, the polarities of the leader spots in each hemisphere again corresponded to those of the polar fields of that hemisphere.

Although Hale had recognized that, in addition to the magnetic fields of sunspots, the Sun possessed a global magnetic field, this was the first indication that the global poloidal field was oscillatory, with a half-period equal to the sunspot number period. If the poloidal field were to complete a full cycle and return to its original configuration after 22 years, the global field periodicity would be the same as the sunspot magnetic periodicity, although $\sim 90°$ out of phase, the poloidal field peaking at sunspot minimum. The inference that the two are causally related was hard to resist, and observers at Mount Wilson and other observatories followed the polar fields with some interest during the next sunspot maximum. It is now history that similar reversals were duly observed in Cycles 20 and 21, and it would appear that they are again (1992) taking place in Cycle 22.

It is perhaps important to stress that, although these observations confirmed that the sunspot magnetic cycle and the global magnetic cycle have the same periodicity, the inference that they are causally connected is just that, an inference, which, although it may seem eminently reasonable, should be subject to the same scrutiny as any other inference concerning the mechanism of the sunspot cycle. It is also taken for granted at present that the polar fields reverse sign around the time of solar maximum. We should bear in mind, however, that we have only three, now almost four, instances on which to base this notion, and a similar caution should be observed.

2.12 Conclusion

As the expansion of Los Angeles and of the automobile industry continued into the 1930s, the unique quality of the Mount Wilson Observatory declined, and solar observations were expanded to other sites, both in the United States (e.g. the Sacramento Peak Observatory in New Mexico and the McMath Solar Telescope at the Kitt Peak Observatory, near Tucson, Arizona) and in Europe (e.g. the Pic du Midi Observatory in the Pyrenees and the Crimea Observatory). Rather than attempting a chronological account of these developments, the next chapter offers a summary of the current background of our knowledge of solar physics as required for an understanding of the cyclic phenomenon.

References

Bray, R. J., and Loughhead, R. E.: 1964, *Sunspots*, Chapman and Hall, London.

Cowling, T. G.: 1953, in *The Sun*, ed. G.P. Kuiper, University of Chicago Press, Chicago, Illinois, 565.

Drake, S.: 1957, *Discoveries and Opinions of Galileo*, New York.

Drake, S.: 1990, *Galileo: Pioneer Scientist*, University of Toronto Press, Toronto.

Eddy, J. A.: 1977, in *The Solar Output and its Variation*, ed. O. R. White, Colorado University Press, Boulder, Colorado, 51.

Hale, G. E.: 1926, *Natural History* **26**, 363.

Hale, G. E., and Nicholson, S. B.: 1938, *Magnetic Observations of Sunspots, 1917–1924*, (Carnegie Institution of Washington), Part I, 56.

Pepin, R. O., Eddy, J. A., and Merrill, R. B.: 1980, *The Ancient Sun: Fossil Record in the Earth, Moon and Meteorites*, Pergamon Press, New York.

Wright, H., Warnow, J. N., and Weiner, C.: 1972, *The Legacy of George Ellery Hale*, MIT Press Media Department.

3

The structure of the Sun and the phenomena of activity

Far out in the uncharted waters of the unfashionable end of the western spiral arm of the Galaxy lies a small, unregarded, yellow sun.

D. Adams, The Hitch-hikers Guide to the Galaxy

O Sole Mio
Neapolitan folk song

3.1 Basic data

The historical studies traced in the previous chapter provided an introduction to our knowledge of the structure of the Sun and of cyclic activity. We now offer a brief summary of the general state of our knowledge of the physical properties of the Sun and of solar-type stars, together with some basic theory relevant to an understanding of cyclic phenomena. The interested reader who desires further information is referred to the more general accounts listed in the references (e.g. Mihalas 1978, Foukal 1990, Stix 1989, Zirin 1989).

Stars are generally classified according to their luminosity and surface temperature, a classification scheme which has been codified as the Hertzprung–Russell (H–R) diagram. In this diagram the absolute magnitude (or logarithm of the total luminosity) is plotted against the logarithm of the surface temperature. In the Harvard classification scheme the categories O, B, A, F, G, K, M, R, and S represent decreasing surface temperatures and increasingly complex spectra, and the Sun (of type G2) sits squarely in the middle. It is about 4.5×10^9 years old, less than half the age of the oldest stars in our galaxy, and in these, as in many other respects, it is a very 'ordinary' star.

As a star contracts gravitationally, it moves downwards (i.e. to lower luminosities) and to the left (to higher temperatures) on the H–R diagram, until its internal temperature reaches a value $\sim 10^6$ K, when the fusion of hydrogen to helium begins. At this stage, the star is said to have reached the *main sequence*, a broad band of stars on the H–R diagram extending from upper left to lower right and occupied, apparently, by stars which are still in the hydrogen-burning stage. The Sun is also located very much in the middle of this range.

The mode of transport by which energy released in the core of a star is brought to the surface depends on the thermal structure of the interior. Although it cannot be observed directly, the internal structure of a star may be modelled by solving the equations of stellar structure, using the observed (or inferred) surface temperature, luminosity, mass, and chemical composition as constraints. (see e.g. Chandrasekhar 1960, Kippenhahn and Weigert 1990).

If the magnitude of the actual temperature gradient at any level is less than that of the local adiabatic gradient, the gas at this level is stable against thermal instabilities, since, if an element of the gas at the ambient temperature of the level rises through a height dz, it cools adiabatically and, being now cooler than the gas at this new level, tends to sink back to its original level. In such a region the transfer of energy is by radiation, the rate determined by the local temperature gradient.

If the magnitude of the actual temperature gradient exceeds the adiabatic gradient, however, a rising element is hotter than its surroundings and continues to rise. Such a region is said to be *convectively unstable*, and the hotter rising gases carry the excess energy upwards. Eventually, the upward motions are checked, either by a restoration of the sub-adiabatic gradient or by a sharp change in density, such as that at the surface; the element gives up its excess energy by heat exchange with its surroundings or by surface radiation and, now cooler than its surroundings, falls back. Under these conditions the star's energy is said to be transported by *convection*, and the temperature gradient is forced to correspond to the adiabatic (or isentropic) gradient.

Solar models indicate that the hydrogen-burning core occupies roughly the inner third of the Sun's radius; radiation transports the energy across the next third and convection across the outer third ($\sim 200\,000$ km) of the radius. This latter region is known as the *convection zone*, and it terminates just below the *photosphere*, the Sun's deepest visible layer, where radiation again takes over. These different layers are illustrated in Figure 3.1 (due to Wilson *et al.* 1981).

Conditions vary considerably in stellar interiors. The earlier type stars (types O, B, A) are almost entirely radiative; shallow convection zones are found in the F-type stars, and the depth of these zones increases through the G, K, and M-type stars, the latter being almost entirely convective. As

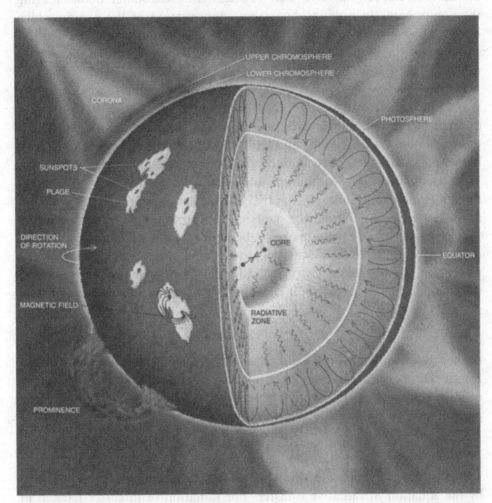

Fig. 3.1 The basic features of solar structures and modes of energy transport are shown in this cross-section of the solar interior. The wavy lines correspond to radiative transport of energy, whereas the looped lines denote the region of convective transport (from Wilson *et al.* 1981).

will be demonstrated in Chapter 7, these structural differences are relevant to the cyclic behaviour of stars.

3.2 Radiative transfer

The photosphere of the Sun (or of a star) is that layer of the atmosphere which can be seen in the visible portion of the continuous radiation spectrum. More precisely, it is that range of depths where photons in the continuum of the visible spectrum have their last scattering encounter before leaving the atmosphere. In order to have some appreciation of the way in which observations of the photosphere can provide information about the physical structure of the atmosphere, some elementary definitions from the field of radiative transfer are required.

Let the opacity, κ_v, be that fraction of a beam of radiation of frequency v and intensity I_v which is absorbed or scattered out of the beam per unit distance by the atoms, molecules, or electrons of the plasma through which it passes. Naturally, κ_v depends on the density of the plasma, being very small in the thin outer layers of the photosphere, but increasing rapidly with depth. The opacity also varies with temperature and with the frequency of incident radiation, increasing sharply from that at the continuum frequency v to that at the center v_0 of an atomic or molecular absorption (i.e. Fraunhofer) line.

A given thickness of plasma may therefore present different opaqueness characteristics to incident radiation, depending on the properties of the plasma and on the frequency of the radiation. For an element of plasma of thickness dz and opacity $\kappa_v(z)$ it is convenient to define an element of *optical thickness* $d\tau$ by

$$d\tau = -\kappa_v(z)dz, \tag{3.1}$$

the negative sign indicating that dz has been defined to be positive outwards. Hence the *optical depth* $\tau_v(z)$ of a point at depth z below the (arbitrarily defined) surface of a stellar atmosphere, or the optical thickness of a plane parallel layer of physical thickness z, is given by

$$\tau_v(z) = -\int_0^z \kappa_v(z)dz. \tag{3.2}$$

The transfer of radiation through the atmosphere of a star is governed by the equation of radiative transfer. For radiation of intensity $I_v(\tau_v, \mu)$ in direction $\mu = \cos\theta$, where θ is the angle between the direction of the beam of radiation and the outward normal, at a depth $\tau_v(z)$ in a plane-parallel atmosphere which is locally in thermodynamic equilibrium at temperature $T(z)$, this equation may be written (see e.g. Foukal 1990, p. 44)

$$\mu\frac{\partial I_v(\tau_v, \mu)}{\partial \tau_v} = B_v(T) - I_v(\tau_v, \mu), \tag{3.3}$$

where $B_v(T)$ is the Planck function at temperature T, i.e.

$$B_v(T) = \frac{2hv^3}{c^2}(e^{hv/kT} - 1)^{-1}. \tag{3.4}$$

For an atmosphere which is not in local thermodynamic equilibrium, the Planck function must be replaced by a more general *source function* S_v.

An elementary solution of this equation yields the intensity, $I_v(\mu)$ $\{=$ $I_v(0, \mu)\}$ of the radiation emerging in direction μ from a plane-parallel semi-infinite atmosphere, which is

$$I_v(\mu) = \int_0^\infty B_v(T)e^{-\frac{\tau_v}{\mu}} \frac{d\tau_v}{\mu}. \qquad (3.5)$$

The Planck function is a measure of the emissivity of radiation per unit volume and solid angle, and must increase with depth as the temperature increases. Further, an approximate form of the equation of radiative transfer (due to Eddington) has an elementary solution of the form

$$B_v = C + D\tau_v, \qquad (3.6)$$

under the assumption of radiative equilibrium (i.e. that all the radiation absorbed by an elementary volume is re-emitted). Substituting in Equation (3.5) yields

$$I_v = C + D\mu, \qquad (3.7)$$

which implies that the energy radiated from the surface at angle $\cos^{-1}\mu$ is equal to the value of B_v when $\tau_v = \mu$ and, in particular, the intensity of the normally emerging radiation corresponds to the value of B_v when $\tau_v = 1$.

As a useful generalization, the physical depth z corresponding to optical depth $\tau_v(z) = 1$ is said to be the origin of the emergent radiation of frequency v, since it is that depth at which the local value of the Planck function corresponds to the intensity of the normally emerging radiation. At frequency v we 'see' into the atmosphere to a depth corresponding to unit optical depth at that frequency, and we may sample the photosphere at different levels by observing at different wavelengths in the continuum. The deepest penetration is obtained at infra-red wavelengths; alternatively, we may sample the higher layers by observing at the centres of the many spectral lines, not only in the visible range of wavelengths, but also in the infra-red and in the ultra-violet.

In the same way, spectral lines of frequency v_0 are said to be formed at the depth corresponding to $\tau_{v_0} = 1$, and a plane-parallel layer of optical thickness τ_v is said to be optically thin at frequency v if $\tau_v \ll 1$, and optically thick if $\tau_v \geq 1$.

The very considerable range of opacities of these lines permits the investigation of the properties of many levels of the atmosphere. In particular, the identification of the normally emerging radiation, I_v, with the Planck function $B_v(T)$ at the depth corresponding to $\tau_v(z) = 1$, permits the association of a temperature with this level. From Equation (3.4) it can be seen that the frequency dependence of $B_v(T)$ will be different for different values of T, and the surface (or radiation) temperature of a star is defined as that temperature at which $B_v(T)$ best matches the emergent radiation spectrum I_v, either by adjusting T so that the maximum of $B_v(T)$ coincides with that of I_v, or by a least-squares fit at several wavelengths. For the Sun I_v peaks near the wavelength 5000 Å, corresponding to a radiation temperature of ~ 5800 K, and this is adopted as the surface temperature of the normal photosphere. At the centre of a large sunspot the corresponding temperature is ~ 4000 K.

The 'surface' or level of the solar atmosphere identified with this temperature is that at which $\tau_{5000} = 1$ (τ_{5000} being the optical depth at the frequency corresponding to a wavelength of 5000 Å). The level cannot, of course, be determined until the opacity κ_v is known at this and all higher levels, and, since κ_v depends not only on density but also on temperature, the development of a self-consistent photospheric model requires many iterations of some basic model. Several models of the solar atmosphere, based on these methods, are now available.

3.3 The limb darkening

On good quality photographs of the full disk of the Sun it is easy to see that the central regions are brighter than the edge of the disk or *limb*. Thus the function $I_v(\mu)/I_v(1)$, where μ is the cosine of the angle between the normal to the solar surface and the direction of the Earth at any point on the solar disk (i.e. $\mu = 1$ at disk-centre and zero at the limb), is called the *limb darkening* and is an increasing function of μ.

For radiation emerging at an angle θ to the normal to a plane-parallel semi-infinite atmosphere, Equation (3.5) may be written

$$\frac{I_v(\mu)}{I_v(1)} = \int_0^\infty \frac{B_v(T)}{I_v(1)} e^{-\frac{\tau_v}{\mu}} \frac{d\tau_v}{\mu}. \qquad (3.8)$$

The inversion of Equation (3.8) provides the most direct means by which an observed quantity, the limb darkening, $I_v(\mu)/I_v(1)$, may provide information about the physical structure (i.e. the temperature distribution) of the solar atmosphere. Since stellar limb darkening cannot yet be reliably measured, this method is not yet applicable to stars other than the Sun.

Although there are mathematical problems associated with this inversion, the power of the method may be illustrated by again using the elementary solution of the Eddington approximation (Equation 3.6), i.e. that the Planck function is a linear function of τ_v,

$$\frac{B_v(\tau_v)}{I_v(1)} = c + d\tau_v.$$

Then substitution in (3.5) yields

$$\frac{I_v(\mu)}{I_v(1)} = c + d\mu. \tag{3.9}$$

Conveniently, a linear representation of $I_v(\mu)/I_v(1)$ provides a fair approximation, and evaluating c and d by a least-squares fit to the observed limb darkening yields a linear representation for the Planck function, and thus the temperature, in terms of the optical depth.

3.4 Departures from spherical symmetry

The one-dimensional radiative transfer theory summarized in §3.2 applies strictly only to spherically symmetric systems. Although the basic atmospheric structure of the Sun and other stars may be inferred by these methods, high-resolution photographs of the solar surface indicate a wealth of structures, including magnetic and velocity field structures, which are clearly important to an understanding of the physical processes which determine the overall configuration of the Sun.

Historically, solar investigations were compartmentalized into studies of the *quiet Sun* and the *active Sun*, a division formalized by the creation of two independent Commissions of the International Astronomical Union (IAU), Commissions X and XII. Today it is widely recognized that this division is artificial, that no regions of the Sun are truly 'quiet' (in the sense that they contain no phenomena of activity) and that one cannot study the active Sun independently of the quiet Sun (and vice-versa), a conclusion reinforced by the almost complete overlap of the membership of the two Commissions of the IAU.

Nevertheless, the term *active region* is in common parlance and obviously describes a region on the Sun in which the phenomena of the active Sun are found. It will therefore be convenient to describe first those departures from spherical symmetry which, together with the one-dimensional description of the photosphere, are traditionally regarded as pertaining to the quiet Sun, and then group together the components of activity which generally comprise an active region.

3.4.1 The quiet Sun

3.4.1.1 The granulation

The earliest high-resolution photographs of the Sun, such as the remarkable photograph obtained by Janssen in Paris in 1885 (see Bray and Loughhead 1967, Plate 1.4), exhibited a fine mottled structure, which has been confirmed by more recent observations and is known as the *photospheric granulation*. A modern example is shown in Figure 3.2. The pattern consists of bright granules of various polygonal shapes separated by narrow dark lanes, the average centre-to-centre spacing of the granules being 1400 km, but with considerable range from over 2000 km down to the resolution limit of ~ 300 km. The intensity contrast between bright and dark features ranges between 10 and 20% at wavelengths around 5500 Å, corresponding to temperature variations of ~ 200 K.

3.4.1.2 The five-minute oscillations

Similarities between the granulation pattern and the appearance of laboratory convection in a thin layer led many to infer that the granulation is evidence of subsurface convection. However, attempts to identify the expected velocity patterns were unsuccessful until it was recognized that the surface velocity field also exhibited a strong oscillatory component, which Leighton, Noyes, and Simon (1962) identified as the *five-minute oscillations* because of their characteristic period. These were eventually interpreted by Ulrich (1970) and Deubner (1972) as standing acoustic waves trapped in resonant cavities below the photosphere. Studies of the rich modal structure of these oscillations, which can be seen when their spectral power is plotted on the plane of spatial wave number k versus temporal frequency ω (known as a k–ω diagram), have been developed into the field of *helioseismology*. This has proved to be a very powerful tool for probing the solar interior and is further discussed in Chapter 12.

When the oscillatory component is filtered out of the velocity data, the residual blue-shifts coincide well with the bright granules and the red-shifts with the dark lanes, which confirms that the granulation is, indeed, convective in origin and, if one assumes aspect ratios based on laboratory convection, originates at depths of order 600 km below the level $\tau_{5000} = 1$.

3.4.1.3 The supergranulation

The techniques developed by Leighton and his colleagues also showed the presence of patterns of outflowing horizontal velocity fields of speeds ~ 0.5 km s^{-1} and dimensions 20 000–40 000 km, which they called the *supergran-*

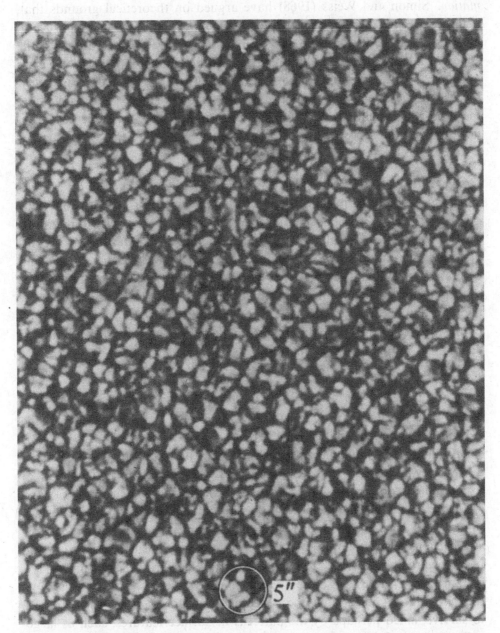

Fig. 3.2 A high-resolution photograph of the photosphere clearly reveals the photospheric granulation consisting of bright granules separated by narrow dark lanes (National Solar Observatory).

ulation. Simon and Weiss (1968) have argued on theoretical grounds that, if the convection zone extends some 200 000 km below the surface, several regimes of cell sizes would be expected, and the surface observations of the supergranulation are generally taken as evidence for the existence of a class of convection cells originating at depths of order 10 000–20 000 km below the photosphere. Attempts to correlate the surface velocity patterns defining the cells with brightness variations appropriate for a convective cell have, however, been unsuccessful.

3.4.1.4 *Giant cells*

Simon and Weiss (1968) have also argued for the existence of a third and larger class of convection cell, which they call the *giant cell*, originating at the base of the convection zone and having transverse dimensions ~ 200 000 km. The nonlinear dynamical studies of convection in a compressible rotating fluid by Gilman and Miller (1986), Glatzmaier (1985), and others indicate that the largest eddies should take the form of 'banana-shaped' cells elongated in the meridional direction with axes (approximately) parallel with the rotation axis. However, Eltayeb (1972) has shown that, in the presence of a toroidal magnetic field, the mode of convection is determined by the Elsasser number

$$\lambda = \frac{B^2}{2\mu\rho\eta\Omega},$$

where B is the intensity of the toroidal field, μ is the permeability, ρ the plasma density, η the magnetic diffusivity, and Ω the angular velocity. For small values of λ, banana cells are favoured, but, for values in excess of a critical value (~ 1), the onset of convection should be in the form of azimuthal (or 'doughnut-shaped') rolls symmetric about the rotation axis.

Such cells might not be expected to extend in coherent form to the surface (Simon and Weiss have argued that any cell might extend only over three scale heights) and attempts to detect giant cells in the surface velocity field (e.g. Howard and LaBonte 1980, Snodgrass and Howard 1984, 1985a,b) have, so far, been unsuccessful. Simon and Weiss have recently (1991) withdrawn their original argument, but the concept of a class of giant cell continues to survive, aided perhaps by the patterns of surface features, such as large-scale magnetic fields and the complexes of activity having the appropriate dimensions.

3.4.1.5 *The differential rotation*

The Sun has a mean rotation period of 26.5 days but, as Carrington discovered, its surface rotates differentially, the equatorial rotation period being

~ 25 days, while at high latitudes the period is ~ 35 days. Although this is believed to be a consequence of the Coriolis effect of rotation on rising and falling convective motions, the mechanism is still not understood. Carrington's inference was made by observing the rotation rates of sunspots at different latitudes, but more recent measurements have been made by studying the motions of other features, or tracers, such as the magnetic field or the supergranule patterns, and by direct observations of the Doppler shifts in spectral lines at various points on the disk.

Although all of these methods confirm that the Sun's atmosphere rotates differentially, they show systematic differences, the rotation rates derived from tracers being uniformly faster than those from spectroscopic data. The tracers include sunspots, plages and other activity-related phenomena, as well as more general surface features, such as the supergranule pattern. Of these, the supergranule pattern yields the fastest rotation rate. Since it is believed that sunspots are 'anchored' to deeper layers of the convection zone by magnetic 'flux ropes' extending below them, while the supergranule cells probably extend to depths of ~ 20 000 km, these results have been interpreted as indicating that the solar rotation rate varies with depth as well as latitude. The question of the internal rotation rates is further discussed in Chapter 12.

Superimposed on the *differential rotation* pattern is a much weaker large-scale velocity pattern, known as the *torsional oscillations*. This term denotes latitude zones of slightly faster- and slightly slower-than-average rotation rates with respect to the local differential rate, which were discovered by Howard and LaBonte (1980) during an unsuccessful search for giant cells. Although the signal is barely above the noise level (~ 6 m s^{-1} compared with the equatorial rotation rate of ~ 2 km s^{-1}), the bands progress from high latitudes to the equator and, when they reach sunspot latitudes, march in synchrony with the butterfly diagram and thus may bear a significant relationship to the phenomena of the activity cycle (see below).

Snodgrass and Wilson (1987) argue that the torsional oscillations represent the surface signature of a deep-seated pattern of doughnut-shaped, azimuthally symmetric convective rolls, which propagate towards the equator. They reason that the subduction zones attract equatorward flows from higher latitudes and poleward flows from lower latitudes which, through the Coriolis effect, would create the shear zone associated with the torsional pattern.

3.4.2 The active Sun

The term *active region* generally describes an area of the solar surface within which many of the phenomena of activity described below are found. Major active regions may extend over areas with linear dimensions $\sim 100\,000$ km and lifetimes of several months, but the term is also applied to regions with dimensions ranging across the whole spectrum of sizes, down to *ephemeral active regions* with sizes $\sim 10\,000$ km and lifetimes of ~ 1 day.

3.4.2.1 Sunspots

For larger active regions, an essential ingredient is at least one, and usually several, *sunspots*. These provide the most dramatic examples of activity and of the non-uniform structure of the photosphere and have already been discussed at some length. Here we briefly summarize their main properties. For a more complete account, the interested reader is referred to the standard text (Bray and Loughhead 1964) or to the more recent proceedings of the NATO Workshop on Sunspots (Thomas and Weiss 1992).

A typical sunspot observed in white light consists of a dark central area, called the *umbra*, surrounded by a roughly annular region consisting of alternately bright and dark approximately radial filaments, called the *penumbra*. A spectacular example is shown in Figure 3.3. Sunspots vary widely in their configurations. Even large spots can be found without penumbrae, while isolated regions of penumbrae can be found away from spots.

The umbral diameter is typically $10\,000$ km, but that of the largest spots may exceed $20\,000$ km. Penumbral widths may be $10\,000$–$15\,000$ km. Sunspots tend to appear in groups (see §2.9 above). These may be relatively simple, consisting of a large *leader* (i.e. westward) spot and one or more *follower* spots, or they may be very complex. As Hale discovered to his surprise, sunspots are associated with magnetic fields of the order of 3000 G, and leader and follower spots are of opposite magnetic polarities, the magnetic bipolar axis (from follower to leader) being generally inclined towards the equator at an angle of $\sim 12°$.

They may also occur in larger groups of varying complexity. Lifetimes of sunspots may range from a few days to several solar rotations. Sunspot groups tend to emerge either sequentially at the same or similar Carrington longitudes, known as *active longitudes*, or to overlap in *clusters*. Small dark structures with diameters < 2500 km and lacking penumbrae are called *pores*. These may exist within groups or may appear as isolated structures, and their lifetimes extend from a few hours to several days. An example of

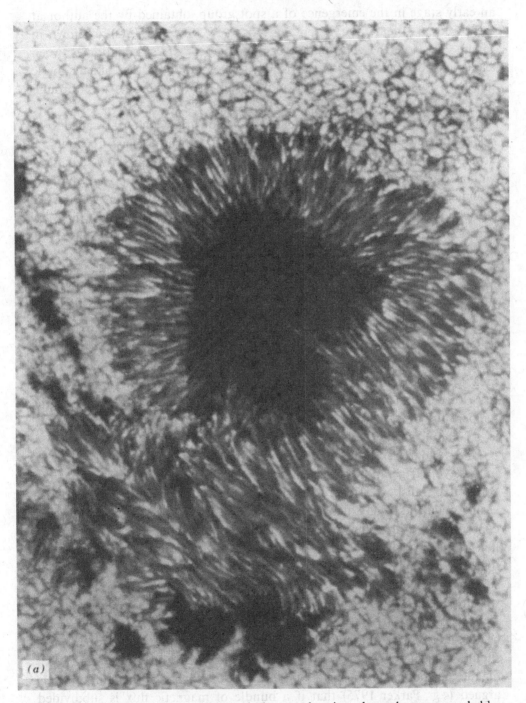

(a)

Fig. 3.3 A typical sunspot exhibits a dark central region, the umbra, surrounded by an annular region, consisting of alternately bright and dark radial filaments, called the penumbra (Observatoires du Pic-du-Midi et de Toulouse).

an early stage in the emergence of a spot group, obtained by the author at the National Solar Observatory, Sunspot, New Mexico (NSO), is shown in Figure 3.4. Note the deformation or alignment of the granulation between leader and follower spots, which strongly suggests that the apex of a magnetic flux loop connecting leader and follower spots is thrusting its way through the photosphere.

The structure of some spots includes lanes of bright material, called *light bridges*, intruding into or traversing the umbra, which may last for hours or days. High-resolution photographs with sufficient exposure show structures within some spots, which were first called *umbral granules* and were thought to indicate that the kilogauss fields of the umbra had not entirely suppressed convection. It is now realized that, unlike photospheric granules, they are aggregations of even smaller structures, called *umbral dots*, with diameters of only a few hundred kilometres and lifetimes of only a few tens of minutes. How such small intense features can exist for these times within the cool umbra is not yet understood.

The energy balance between spots and their surroundings is of importance not only to an understanding of the cyclic phenomenon but also to our terrestrial environment, which may respond in non-linear fashion to small changes in the solar output. Stefan's law that the total energy emitted per unit area by a black body at temperature T is proportional to T^4 implies that, since the umbral temperature is ~ 4000 K compared with a photospheric temperature of ~ 5800 K, the energy flux through a given area of the umbra is $\sim 20\%$ of that through an equivalent area of the photosphere.

The standard explanation, originally due to Biermann (1941), of this impressive refrigeration of a significant area of the photosphere is that the convective transport of energy below the spot is inhibited by its underlying strong magnetic field. Calculations show that, if the field below a spot were concentrated so that convection was inhibited altogether, the surface energy flux would be reduced to only $\sim 5\%$ of the photospheric flux. This has led to the *cluster* model (Parker 1979) in which the sub-umbral magnetic field structure is divided into bundles of flux, separated by non-magnetic regions, within which some form of modified convection may occur.

The situation is, however, far from clear. When Parker (1975) made his challenging remark that 'sunspots should never form and, if once formed, should immediately break apart', he was concerned about the likely effect on the sunspot magnetic field of the 'flute' or 'exchange' instability. It is argued (e.g. Parker 1975) that if a bundle of magnetic flux is subdivided into a number of smaller bundles, the total magnetic energy is unchanged and therefore the stability is neutral. However, if there is a constriction or

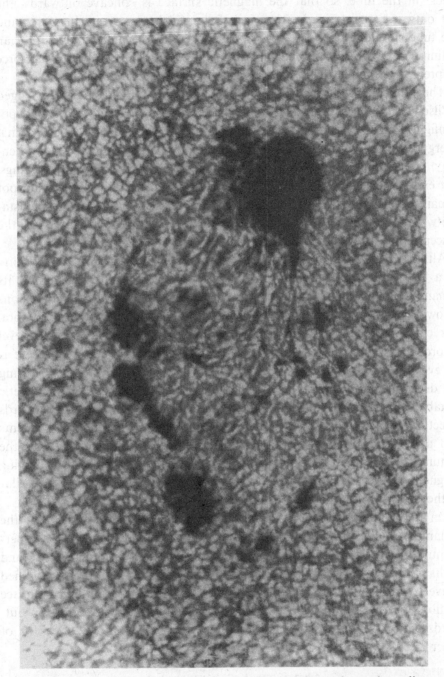

Fig. 3.4 An early stage in the emergence of a sunspot group shows the rudimentary leader and follower spots. Between these an alignment of the granules suggests the emergence of the top of a magnetic field loop (National Solar Observatory).

'neck' in the tube, so that the magnetic surface is concave outwards, the net outward tension in the tube is unbalanced by any restoring force and the tube should subdivide. Thus if 'fluting' (after the manner of a Grecian column) occurs in the curved surface of the tube there is no restoring force to prevent the 'flutes' from deepening until the tube subdivides.

The applicability of this result to the sunspot flux tube has been challenged (Wilson 1977) on the grounds that, when the total energy of the sunspot configuration is considered (including both the potential and the thermal energy), the subdivision process is endothermic (i.e. it requires external energy) if the temperature in the tube is sufficiently less than its surroundings, but exothermic if the temperatures are comparable. Thus, while a sunspot remains cooler than its surroundings, it should survive this particular instability. Meyer *et al.* (1977) have also used the energy principle to assess the susceptibility of the tube to surface fluting and obtained a similar result.

Although these arguments appear to satisfy concerns regarding the stability of a 'static' sunspot magnetic field, Parker was also concerned about its dynamic stability, and his cluster model required a downflow in regions below the umbra in order to maintain the dynamic stability of the spot. For continuity, this requires a radial inflow into the spot at some level in order to match the downflow but, as Parker readily admitted, there is no evidence of such a flow, which would be particularly important during the decaying phase when the spot should be most vulnerable to magnetic instabilities. Such systematic flows that are observed are in the outwards direction: the *Evershed* flow of ~ 5 km s^{-1} along the penumbral filaments and a photospheric flow of ~ 0.5 km s^{-1} across the region beyond the penumbra, called the *moat*, which is accompanied by outwardly moving magnetic features. These flows across the moat have characteristics similar to those associated with the supergranules.

On the evidence of these flows, Meyer *et al.* (1974) proposed that the dynamic stability of a spot is maintained by a rearrangement of the supergranule cells around it, so that the upflows of the cells are adjacent to (and perhaps interweave with) the magnetic structure below it. Some modified convective transport is thus possible below the spot. The observed surface outflows are generated by these flows, which drag magnetic elements outwards across the moat, and also entail a subsurface inflow, at depths of order 20 000 km, which stabilizes the magnetic configuration.

3.4.2.2 *The magnetic network*

Figure 3.5(a) shows a full-disk magnetogram obtained at the Kitt Peak Station of the National Solar Observatory on 23 June 1980, i.e. near sunspot

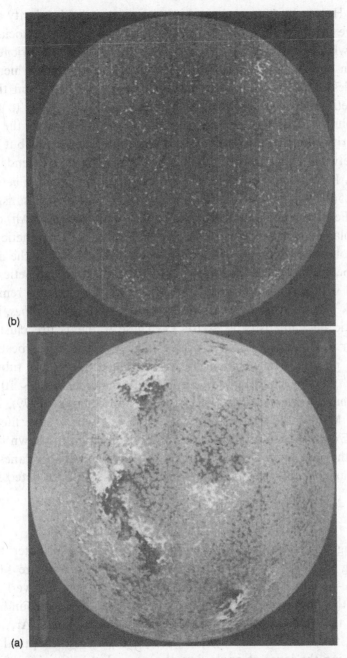

Fig. 3.5 The magnetic network of the Sun is revealed in magnetograms which indicate the polarities of the photospheric magnetic field by a grey scale; white regions represent a region of positive field and black represent negative fields. (a) A full-disk magnetogram obtained at Kitt Peak Station of the National Solar Observatory on 23 June 1980, near sunspot maximum. (b) A full-disk magnetogram taken near sunspot minimum, 28 February 1977.

maximum. Here, white represents regions of positive field polarity and dark the negative field. In addition to the strong bipolar regions associated with sunspots (which illustrate the Hale–Nicholson law for the orientation of polarities in opposite hemispheres), this magnetogram shows a hierarchy of spatial scales in the photospheric magnetic fields, ranging from the large-scale (sometimes carelessly called the 'unipolar') fields, down to the limit of the resolution of the magnetograph (~ 1 arcsec). However, the fields of either polarity outside sunspots are not distributed smoothly but form an irregular network pattern called the *magnetic network* which extends over the whole disk. Figure 3.5(a) provides a good example of the *active network*.

Figure 3.5(b) is a similar full-disk magnetogram taken near sunspot minimum (28 February 1977) and is typical of the *quiet network*. Although the strong bipolar regions are absent, the network consists of magnetic features of mixed polarities which are related to the small-scale end of the active region size spectrum. Careful analyses have shown that the magnetic network is cospatial with the boundaries of the supergranule cells, but, remarkably, the network fields consist of small, discrete elements, or *flux tubes*, in which the magnetic field intensity is $\sim 10^3$ G. These elements are clustered together on a wide range of spatial scales and separated by, at best, weakly magnetized plasma. The magnetic fluxes contained in the flux tubes range from $\sim 10^{19}$ Mx for small bipolar regions with major axes of $\sim 10\,000$ km, called *ephemeral active regions* (ERs) (Martin and Harvey 1979), down to $\sim 5 \times 10^{17}$ Mx for the smallest network elements, which have lifetimes of ~ 20 min. Examples can be seen in the NSO magnetogram shown in Figure 3.6, where they are highlighted by circles. Beyond the network lanes smaller magnetic flux elements, called intranetwork fields, have been detected.

3.4.2.3 *Photospheric faculae*

On white-light photographs of the solar disk, regions brighter than the surrounding photosphere can be found near the limb and are known as *photospheric faculae*. It is argued that these are structures overlying and parallel to the photosphere, which are hotter than their surroundings. At disk centre they are invisible because their optical thickness, $\Delta\tau_v$, is much less than 1, but, near the limb where $\mu = \cos\theta << 1$, the optical path of light traversing the layer at angle θ to the normal, $\Delta\tau_v/\mu$, may be of order 1, and they become visible. Away from active regions the faculae are weak and appear to be closely aligned with the elements of the quiet magnetic network. In the neighbourhood of sunspots, where the network is more densely packed, they tend to overlap and can be identified further from the

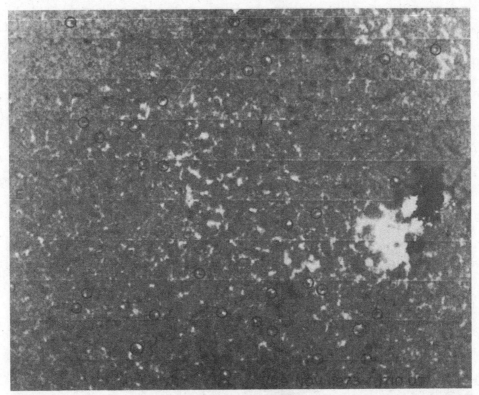

Fig. 3.6 An NSO magnetogram, taken on 29 November 1973, shows small magnetic network elements which are highlighted by circles.

limb. An example of faculae associated with a sunspot near the limb is shown in Figure 3.7.

Photospheric faculae appear in increased numbers in a region prior to the emergence of sunspots and remain for a rotation or more after the spots have decayed. Because they are brighter than their surroundings, they are believed to be hotter and thus to radiate more energy. The faculae are, for this reason, important to the energy balance between sunspots and the photosphere.

3.4.2.4 Filigree

The smallest structures which have been resolved on the disk are the *filigree*, discovered by Dunn and Zirker (1973) and shown in Figure 3.8. Dunn and Zirker noticed that, at low resolution, the granulation pattern appeared 'fuzzy', but, under conditions of excellent seeing, they discovered that this fuzziness is due to small structures (as small as 150 km) in the intergranular

Fig. 3.7 The regions which are brighter than the surrounding photosphere close to the limb are the photospheric faculae. It is argued that these structures are hotter than their surroundings and are closely aligned to the magnetic network away from the active regions (National Solar Observatory).

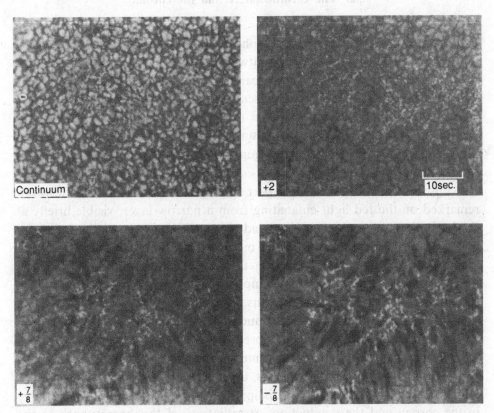

Fig. 3.8 The filigree structures are the smallest that have been resolved on the solar disk. The four photographs were taken at different wavelengths, corresponding to different layers in the photosphere. Figure 3.8(a) was taken at the continuum wavelength with subsequent figures at a wavelength which is offset to the continuum by the value in the bottom left corner in angstroms (National Solar Observatory).

lanes, which can best be seen in the far wings of lines such as MgIb and Hα. These structures are cospatial with, but much smaller than, the faculae and are believed to delineate the magnetic field network associated with faculae.

Active regions are not merely two-dimensional phenomena confined to the photosphere. Many features of active regions extend above the photosphere to higher levels of the solar atmosphere, where they are essential components of the regions known as the *chromosphere* and the *corona*. The magnetic fields associated with an active region must also extend below the photosphere, and the sub-surface configuration of these fields is a fundamental problem of activity which will be discussed in later chapters.

3.5 The chromosphere and the corona

The central temperature of the Sun is $\sim 10^6$ K, while the surface brightness temperature is ~ 5800 K. Above the photosphere, the laws of thermodynamics would seem to require a further outwards decline in the temperature since, according to the second law, heat 'cannot flow up hill'. This is consistent with the limb darkening and the presence of dark (absorption) spectral lines which, following the argument given in §3.2 above, are formed at various heights above $\tau_{5000} = 1$. On the assumption that the regions of formation are in local thermodynamic equilibrium (LTE), the lower central intensities imply lower temperatures.

Eclipse watchers in the latter part of the nineteenth century, however, remarked on the red light emanating from a narrow layer, visible briefly at the beginning and end of totality, and they called this the *chromosphere* or 'colour-sphere', as distinct from the more brilliant *corona*. The subsequent identification of the 'forbidden' coronal lines with those of Fe XIV and XV implied the existence of coronal temperatures of $\sim 10^6$ K and forced the recognition that the feeble radiations from the chromosphere and corona are emitted by plasmas which are much hotter than the underlying photosphere. The theoretical problems posed by this remarkable reversal of the temperature gradient in the Sun's outer atmosphere, in apparent disregard of the basic laws of thermodynamics, are with us today, but the mechanism is believed to be related to the dissipation of hydromagnetic wave energy in the corona and the conduction, or radiation, of this energy back to the chromosphere.

Although originally thought to be a peculiar property of the Sun, chromospheres and coronae are now believed to exist on many stars of the G, K, and M classes (see Chapter 7).

3.6 The chromospheric structures

Apart from eclipse observations, the chromosphere can be observed in spectral lines, with line centre ν_0, for which the level $\tau_{\nu_0} = 1$ is located above the *temperature minimum*, the level at which the temperature begins its outward increase. The most commonly used lines are the hydrogen line $H\alpha$, in the red region of the spectrum (hence the red glow seen at eclipse), and the H and K lines of ionized calcium (CaII), which are located in the violet region. Other useful lines are $H\beta$, $H\gamma$, the H and K lines of MgII, and the 10830 line of helium. Because these lines are formed at rarefied levels at which the assumption of LTE is not valid, the general source function S_ν must be used in place of the Planck function in Equation (3.5), and each line provides slightly different information about the region as a result of the physical properties which determine its source function (see e.g. Cannon 1985).

The chromosphere can be observed either at, or just beyond, the photospheric limb, where these lines are seen in emission against a dark background, or on the disk, where they appear in absorption. The Hα line has a central intensity of a few per cent of the local continuum, but the CaII H and K lines exhibit self-reversed emission peaks at the line centres, which are consistent with the rise in temperature above the temperature minimum.

The chromosphere begins just above the temperature minimum region at the limb, about 500 km above the level of $\tau_{5000} = 1$. In Hα it appears to consist of elongated, tilted structures, known as *spicules*, extending typically ~ 7000 km above the limb. These are not static structures, having lifetimes between 5 and 10 min. Elongation speeds of ~ 25 km s^{-1} have been measured.

Some very different patterns appear on the disk. Figure 3.9 shows parallel images of the full disk of the Sun (a) in white light (continuum radiation), (b) in Hα, and (c) in CaII K. Two parallel rows of sunspots can be readily identified in the white-light image, but in both the Hα and the CaII K photographs they are rendered almost invisible by the bright structures surrounding them.

These bright regions are known as *plages*, from the description provided by French observers, 'plage (beach) faculaire', and are the extensions into the chromosphere of the magnetic structures which were identified with the photospheric faculae. The latitude bands of 'activity' associated with sunspots in each hemisphere are very obvious. The dark irregular lines which appear to meander through the plages are simply prominences observed in absorption on the disk, where they are known as *filaments*.

Away from the plage regions, the Hα chromosphere appears to consist of dark filamentary *mottles*, which are probably the same structures as the spicules seen in emission at the limb, and are believed to delineate magnetic field lines arching into the chromosphere.

Although there are some similarities, the full-disk photograph taken in CaII K at the same time is strikingly different in several respects. The bright plage regions are there, but, in place of the dark mottled filamentary structures, the main impression is of a bright network surrounding darker island structures. The pattern, known as the *chromospheric network*, is the 'photographic negative' of the granulation pattern, but on a larger scale, the typical scale size of the network being 10 000–20 000 km, which is the supergranule scale size. The bright network has been shown to be

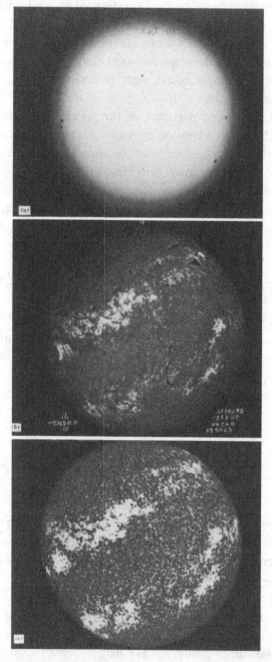

Fig. 3.9 Full-disk images of the Sun were obtained in (a) white light (continuum radiation), (b) Hα and (c) CaII K. Two parallel rows of sunspots are visible in the white-light image; however, they are invisible in the other two images due to the bright structures known as plages, which are extensions into the chromosphere of the photospheric magnetic structures (National Solar Observatory).

Fig. 3.10 A quiescent prominence exhibits prominent vertical patterns due to plasma motion towards the photosphere (National Solar Observatory).

cospatial with the magnetic network, which is identified above with the supergranulation structures, so these two phenomena are clearly related and have been modelled by Stein and Nordlund (1989).

3.7 Prominences

The dark filamentary structures seen against the disk in Hα photographs, such as Figure 3.9(b), take on a most dramatic appearance when photographed beyond the limb against the darker coronal background. They appear as pinkish clouds above the chromosphere at eclipse but are also visible in narrow-band Hα filtergrams, extending to 50 000 km or more above the limb. Figure 3.10 shows an example of a *quiescent prominence* exhibiting extraordinary, predominantly vertical, patterns along which plasma appears to drain towards the photosphere.

Figure 3.11 shows a spectacular example of an *eruptive prominence*. In the first image, the prominence extends some 110 000 km (which is ~ 15% of the solar radius) above the photosphere and is rising at speeds of order 160

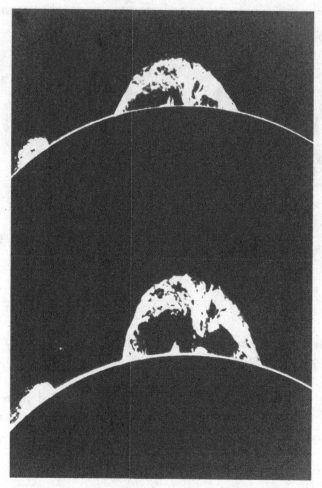

Fig. 3.11 The time evolution of an eruptive prominence is shown in the two images separated by an 11-minute time interval. In the first photograph the prominence extends to a height of 110 000 km above the photosphere, rising with speeds of the order of 160 km s^{-1}. In the second frame, the prominence has attained a height of 370 000 km (Institute for Astronomy, University of Hawaii).

km s^{-1}. In the second image, taken 11 min later, it has reached its full height of 370 000 km (i.e. $\sim 50\%$ of the solar radius).

3.8 Flares

Carrington's observation of a solar *flare* in 1859 (see §2.4) was a fortunate occurrence, because that flare was a particularly intense member of the species, a *white-light flare*. A flare may be defined as a catastrophic and localized release of energy in the form of electromagnetic radiation and leading to particle acceleration. The flare event is perhaps the most spectacular component of solar activity. Flares have also been observed on many classes of stars, and these are discussed in §7.4.

Under normal circumstances, flares are observed in Hα or other chromospheric lines, and white-light flares are a comparative rarity. In Hα, flares are brilliant transient brightenings accompanied by the eruption of material in sprays and surges, together with bursts at radio wavelengths. The most spectacular flare event ever recorded occurred near the west limb of the Sun on 1 and 2 March 1969. It began as a spectacular high-speed prominence ejection observed in Hα at the Mees Solar Observatory in Hawaii, extending several hundred thousand kilometres beyond the limb. Subsequently, the event was recorded in Australia at 80 MHz by the Culgoora radioheliograph as a moving (type IV) burst and was traced out to $5\frac{1}{2}$ solar radii beyond the limb; it became known familiarily as the 'Westward Ho' event.

With the advent of space astronomy, flares are now observed in the EUV and X-ray regions of the spectrum. An example of an Hα flare is shown in Figure 3.12. The Hα brightness may increase up to tenfold, and the line may be broadened by up to 20 Å, suggesting internal velocities up to 1000 km s^{-1}. Particle streams of electrons and nucleons are generally associated with the more intense flares. The impact of these streams on the Earth's ionosphere causes geomagnetic storms, interruptions to telecommunications, and the breakdown of power stations.

The dominant role of magnetic fields in the structure of the corona was established in the late 1960s as high-resolution images became available from broad-band rocket-borne telescopes. Flares are never observed far from active regions, but not all active regions are likely to generate them. Large, almost symmetric sunspots may exist for several rotations but are also unlikely to give rise to flares. Generally, the more complex regions are the likely sites for intense flare activity, particularly those in which the leader and follower spots adopt unusual configurations, which suggests that the subsurface flux ropes are being distorted in some way.

Because flares tend to occur near unusual sunspot configurations, it has been argued that energy is stored in coronal magnetic field structures by the distortion of potential (i.e. minimum energy density) configurations into highly twisted forms. It is assumed that this energy is then released following magnetic reconnections which reduce the field configuration to simpler forms. However, the energy released by a large flare is of the order 10^{32} erg, and how such a large quantity of energy can be released so quickly remains an unsolved problem. To put the problem in perspective, the deficit in the

Fig. 3.12 An example of an Hα flare. Flares are observed as transient brightenings accompanied by the eruption of material in sprays and surges. More intense flares produce particle streams which are known to be the cause of geomagnetic storms upon impact on the Earth's ionosphere (Big Bear Solar Observatory).

energy radiated by a large sunspot compared with that from an equivalent area of the normal photosphere is $\sim 2 \times 10^{29}$ erg s^{-1}, and the energy released by a flare over a few minutes is equivalent to that energy deficit radiated from an equivalent area over a period of 1–2 hours.

Although flares are an important component of stellar activity, an understanding of their physics is still some way off, and the interested reader is referred to the review by Haisch, Strong, and Rodonò (1991), or other references cited in the bibliography, for more detailed information.

3.9 Coronal structures

White-light photographs taken during eclipses show coronal structures down to the smallest resolvable scales (~ 1 arcsec). Artificial eclipses are generated within a *coronagraph* by adjusting an occulting disk at the focal plane of a normal telescope to just block the image of the photosphere and permit lower quality coronal images to be obtained on a regular basis.

Figure 3.13(*a*), obtained at the Indian eclipse of 1980 near sunspot maximum, displays a wealth of coronal structures. Helmet-type structures consist of bright coronal arches with low coronal intensity near their bases and narrow radial streamers extending outward from the top of the 'helmet'. These arch systems appear to indicate closed magnetic structures connecting surface regions of opposite polarities. Other structures consist entirely of radial streamers and may be traced out to 1–2 solar radii from the disk.

Figure 3.13(*b*), obtained in 1954 in Wisconsin near sunspot minimum, shows that the coronal emission is generally reduced at this time and concentrated towards lower latitudes. (The apparently comparable extent of this 'quiet' corona, by comparison with the 'active' corona shown in Figure 3.13(*a*), is due to increased exposure of the photographic plate.) At high latitudes, shorter streamers, called *polar plumes*, can be observed extending out from the poles as if following the field lines of a global bipolar magnetic field.

The higher density structures of the lower corona can be observed at the limb with coronagraphs equipped with narrow-band filters passing the green and red coronal emission lines of FeXIV. Closed loop structures are seen in active regions, while vertical structures, suggesting more open field lines, are evident in the quieter limb regions. Synoptic green-line data obtained with the coronagraph at the Sunspot observatory of the NSO have yielded important data concerning the cycle which will be discussed in Chapter 8.

The appearance of the corona on the disk is best revealed in broad-band soft X-ray photographs, such as those shown in Figure 3.14, in which the

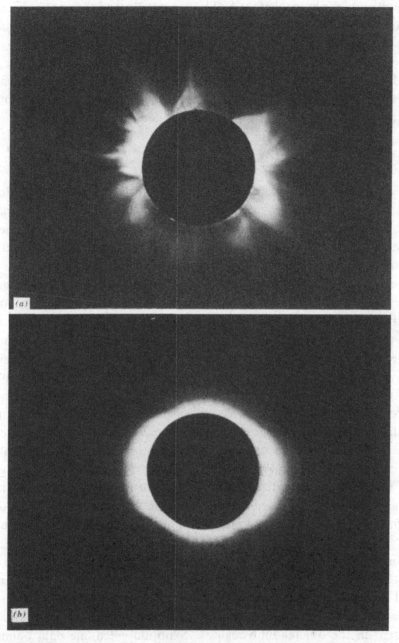

Fig. 3.13 (a) The Indian eclipse of 1980, near sunspot maximum, displays a wealth of coronal structures. (b) In the Wisconsin eclipse of 1954, near sunspot minimum, the coronal emission is reduced, and, at high latitudes, polar plumes (short streamers) are observed extending out from the poles as part of the global magnetic field of the Sun.

most striking features are the very bright active-region condensations and
loops. Fainter, more extensive loop systems linking different active regions
can also be seen.

3.10 Coronal holes

Figure 3.14 also shows the presence of a dark, elongated region, extending
from the northern polar region across the equator into the southern hemisphere, in which the soft X-ray emission is about two or three times lower
than in adjacent structures. There is also a significant decrease in the coronal
green-line emission from such regions, which are now called *coronal holes*,
relative to quiet-Sun values. Coronal holes tend to form near the centres
of large unipolar magnetic regions, and a comparison of the X-ray pictures
with those of magnetic field lines calculated on the assumption that the
observed photospheric field structures extend into the corona as potential
fields (Figure 3.15) indicates that they are regions of open, perhaps diverging, magnetic fields, although their formation seems to require the additional
presence of newly-emerged active-region fields.

Coronal holes may also be observed in spectroheliograms taken in the
10830 line of helium. Unlike smaller surface features, coronal holes tend to
rotate semi-rigidly and more slowly than sunspots or supergranule patterns
(Sheeley *et al.* 1987).

Although coronal holes may form at any latitude, the polar coronal holes
associated with the unipolar coronal fields are of greatest significance to
the cycle; they are well defined at sunspot minimum, when the polar fields
are strongest, and disappear altogether during the polar field reversals near
sunspot maximum. Low-latitude and equatorial coronal holes also play a role
in the activity cycle which is far from understood but clearly of significance,
since intense activity may often be found just beyond their boundaries.

A particularly interesting example of this can be seen in synoptic charts
constructed by McIntosh (1992). Just as the polar fields were beginning their
reversals in 1979, a stable equatorial coronal hole appeared, and, during
the following months, this hole was intermittently attached to the north
polar coronal hole (of like polarity). Significantly, the most outstanding
active region of that year emerged adjacent to this feature at the time of its
maximum development. After this event the polar fields began the reversal
process and no further attachments to the polar hole occurred. In 1982, a
remarkable convergence of three equatorial coronal holes to form the largest
coronal hole of the cycle coincided with the emergence of an active region
at the same longitude, and this region gave rise to high-temperature flares

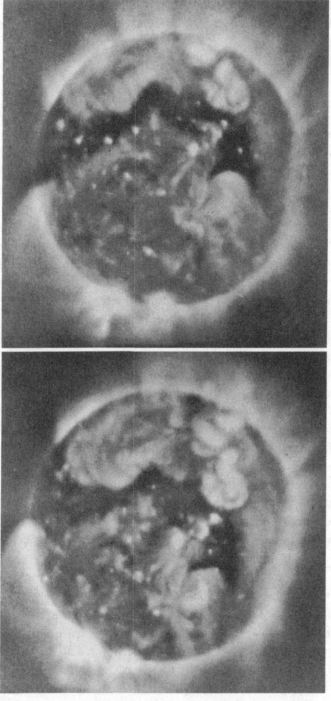

Fig. 3.14 The presence of a coronal hole can be seen in these X-ray photographs as the dark elongated region extending from the northern polar region across the equator into the southern hemisphere.

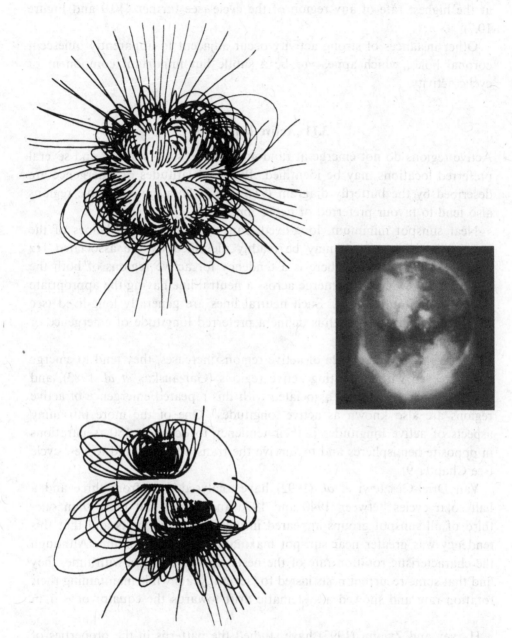

Fig. 3.15 The coronal magnetic field lines have been calculated on the assumption that they are the potential field extension into the corona of the photospheric fields. An X-ray photograph taken at the same time is shown for comparison. It can be seen that the coronal hole is located near the centre of a large unipolar magnetic region.

at the highest rate of any region of the cycle (see further §10.9 and Figure 10.7).

Other instances of strong activity occur adjacent to apparently quiescent coronal holes, which appear to be a subtle but important component of cyclic activity.

3.11 Active longitudes

Active regions do not emerge at random over the solar surface, and several preferred locations may be identified. Preferred latitudes of emergence are described by the butterfly diagram (§2.6 and §8.1), although active regions also tend to favour preferred or *active* longitudes.

Near sunspot minimum, long-lived trans-equatorial neutral lines of the large-scale field patterns may be readily identified by the associated Hα filaments (see §9.2), and there is a tendency for active regions of both the old and the new cycle to emerge across a neutral line having the appropriate polarity orientation (§9.5). Such neutral lines are generally long-lived (see Figures 9.2 and 9.3) and thus define a preferred longitude of emergence or active longitude.

As the rate of emergence of active regions increases, they tend to emerge near, or even within, existing active regions (Gaizauskas *et al.* 1983), and the Carrington longitudes associated with this repeated emergence of active regions are also known as active longitudes. One of the more intriguing aspects of active longitudes is their tendency to occur at 180° separations in opposite hemispheres and to survive the transition from old to new cycle (see Chapter 9).

Van Driel-Gesztelyi *et al.* (1992) have analysed data from three and a half solar cycles between 1940 and 1976 and found that more than one-third of all sunspot groups appeared in compact active nests and that this tendency was greater near sunspot maximum than at minimum. Although the characteristic rotation rate of the nests showed changes with time, they find that some recurrent nests lasted for several years while maintaining their rotation rate and showed a systematic drift towards the equator of ~ 1 m s^{-1}.

Harvey and Zwann (1993) have studied the patterns in the properties of active regions emerging during Cycle 21. They find that, of the regions larger than 3.5 square degrees, 44% emerged within existing sunspotted active regions; thus the rate at which new regions emerge is much higher within the boundaries of existing large active regions than it is in the activity zones outside such active nests.

The concept of an active longitude is not, however, confined to the sunspot latitudes of one hemisphere; examples are given in Chapters 9 and 10 of trans-equatorial active longitudes and of the sequential emergence of an active region at sunspot latitudes and a smaller (i.e. unspotted) region near latitude 50° at the same Carrington longitude.

3.12 Conclusion

These are the solar structures and activity phenomena which take part in the solar activity cycle. In Chapter 7 it will be shown that many of these phenomena have been identified as occurring in solar-type stars, frequently in more extreme, or violent, form. Some features, such as the granulation or the filigree structures, have not yet been identified in other stars, but it would seem that only the technical difficulties of resolving such small features in these distant objects stands in the way of their identification.

These correspondences, which are sometimes referred to as the *solar–stellar connection*, are further explored in Chapter 7.

References

Biermann, L.: 1941, *Vierteljahrsschr. Astr. Gesellsch.*, **76**, 194.

Bray, R. J., and Loughhead, R. E.: 1964, *Sunspots*, Chapman and Hall, London.

Bray, R. J., and Loughhead, R. E.: 1967, *Solar Granulation*, Chapman and Hall, London.

Cannon, C. J.: 1985, *Astrophys. J.*, **289**, 363–72.

Chandrasekhar, S.: 1960, *Radiative Transfer*, Dover, New York.

Deubner, F. L.: 1972, *Solar Phys.*, **22**, 263–75.

Dunn, R. B., and Zirker, J. B.: 1973, *Solar Phys.*, **33**, 281–304.

Eltayeb, I. A.: 1972, *Proc. Roy. Soc.*, **326**, 229–54.

Foukal, P. 1990, *Solar Astrophysics*, John Wiley and Sons, New York.

Gaizauskas, V., Harvey, K. L., Harvey, J. W., and Zwaan, C.: 1983, *Astrophys. J.*, **265**, 1065.

Gilman, P. A., and Miller, J.: 1986, *Ap. J. Suppl.*, **61**, 585.

Glatzmaier, G. A.: 1985, *Astrophys. J.*, **291**, 300.

Haisch, B., Strong, K. T., and Rodonò, M.: (1991) *Ann. Rev. of Astron. and Astrophys.*, **29**.

Harvey, K. L., and Zwaan, C.: 1993, *Solar Phys.* (to appear).

Howard, R., and LaBonte, B. J.: 1980, *Astrophys. J.*, **239**, L33–6.

Kippenhahn, R., and Weigert, A.: 1990, *Stellar Structure and Evolution*, Springer-Verlag, Berlin.

Leighton, R. B., Noyes, R. W., and Simon, G. W.: 1962, *Astrophys. J.*, **135**, 474–99.

Livingstone, W. C.: 1982, *Nature*, **297**, 208–9.

McIntosh, P. S.: 1992, *Solar Phys.*, (in press).

Martin, S. F., and Harvey, K. L.: 1979, *Solar Phys.*, **64**, 93–108.

Meyer, F., Schmidt, H. U., Weiss, N. O., and Wilson, P. R.: 1974, *Mon. Not. R. Astr. Soc.*, **169**, 35.

Meyer, F., Schmidt, H. U., and Weiss, N. O.: 1977, *Mon. Not. R. Astr. Soc.*, **179**, 741.

Mihalas, D.: 1978, *Stellar Atmospheres*, 2nd edn., W. H. Freeman and Company, San Francisco.

Parker, E. N.: 1975, *Solar Phys.*, **40**, 291.

Parker, E. N.: 1979, *Astrophys. J.*, **234**, 333–47.

Sheeley, N. R. Jr., Nash, A. G., and Wang, Y. M.: 1987, *Ap. J.*, **319**, 481.

Simon, G. W., and Weiss, N. O.: 1968, *Z. Astrophys.*, **69**, 435.

Simon, G. W., and Weiss, N. O.: 1991, *Mon. Nat. R. Astr. Soc.*, **252**, short comm's., 1

Snodgrass, H. B., and Howard, R.: 1984, *Ap. J.*, **284**, 848.

Snodgrass, H. B., and Howard, R.: 1985a, *Solar Phys.*, **95**, 221.

Snodgrass, H. B., and Howard, R.: 1985b, *Science*, **228**, 945.

Snodgrass, H. B., and Wilson, P. R.: 1987, *Nature*, **328**, 696.

Stein, R. F., and Nordlund, A.: 1989, *Ap. J.(Letters)*, **342**, L95.

Stix, M., 1989, *The Sun*, Springer-Verlag, Berlin.

Thomas, J. H., and Weiss, N. O.: 1992, *Sunspots: Theory and Observation*, NATO ASI Series, Kluwer, Dordrecht.

Ulrich, R. K.: 1970, *Astrophys. J.*, **162**, 933–1002.

van Driel-Gesztelyi, L., van der Zalm, E. B. J., and Zwaan, C.: 1992, in *Proceedings of the National Solar Observatory/Sacramento Peak 12th Summer Workshop*, ed. K. L. Harvey, San Francisco, California, 89.

Wilson, O. C., Vaughan, A. H., and Mihalas, D.: 1981, *Sci. Am.*, **244**, 89–91.

Wilson, P. R.: 1977, *Solar Physics*, **55**, 35–45.

Wilson, P. R., and McIntosh, P. S.: 1991, *Solar Physics*, **136**, 221–37.

Zirin, H.: 1989, *The Astrophysics of the Sun*, The University Press, Cambridge, UK.

4

The equations of magnetohydrodynamics and magnetohydrostatics

His most aesthetic, very magnetic fancy took this turn–
'If I can wheedle a knife or needle, why not a silver churn?'
W. S. Gilbert, Patience

4.1 Maxwell's equations and the induction equation

Most of the universe is permeated by electrically charged particles and by electromagnetic fields, and the phenomena of activity described in the previous chapter are particular examples of the interactions between these particles and fields which constitute the subject of *cosmic electrodynamics*. It is therefore necessary to set out the basic equations which govern these interactions and to list the approximations and assumptions which underlie our theoretical discussions of the nature of activity and activity cycles.

The motion of a particle of mass m and charge e moving with velocity $\mathbf{v}$ in an electric field $\mathbf{E}$ and magnetic field $\mathbf{B}$ is determined by the equation of motion

$$m\frac{d\mathbf{v}}{dt} = e(\mathbf{E} + \mathbf{v} \times \mathbf{B}). \tag{4.1}$$

While the study of the individual particle motions is of basic importance, the summation of the effects of individual particles is not generally suitable as a method of describing macroscopic solar phenomena, because the combined motions constitute an electric current which changes the electric and magnetic fields. Consequently, it is preferable to derive equations simultaneously relating the macroscopic quantities, the current density $\mathbf{j}$ and the fluid velocity $\mathbf{u}$, to the electric and magnetic fields. This approach defines the *magnetohydrodynamic* or MHD method.

While the general MHD equations tend to be complicated, they may be simplified or idealized by assuming that the mean free path of the

61

particles between collisions is much less than the cyclotron radius mv/Be. The equations discussed in the following section, specifically the induction equation and the equations of fluid dynamics, are known as the *idealized* MHD equations and allow considerable insight into the hydromagnetic processes which underlie the phenomena of solar and stellar activity.

The interactions between the electric and magnetic fields are described by the set of equations attributed to Maxwell and are usually written in mks units. However, astrophysicists tend to use the em unit for magnetic intensity, the *gauss* (G), rather than the appropriate mks unit for induction, the *tesla*, and for the sake of familiarity, if not rigour, this practice will be followed here. The purist will readily make the transition using the relation that, in free space, 1 tesla$= 10^4$ G. Maxwell's equations are

$$\nabla \times \mathbf{B} = \mu\mathbf{j} + \frac{1}{c^2}\frac{\partial \mathbf{E}}{\partial t}, \qquad (4.2)$$

$$\nabla \times \mathbf{E} = -\frac{\partial \mathbf{B}}{\partial t}, \qquad (4.3)$$

$$\nabla \cdot \mathbf{B} = 0, \qquad (4.4)$$

$$\nabla \cdot \mathbf{E} = \frac{q}{\epsilon}. \qquad (4.5)$$

Here $\mathbf{B}$ is the magnetic induction (often loosely referred to as the 'magnetic field' or the 'intensity'), $\mathbf{E}$ is the electric field, $\mathbf{j}$ is the current density, q is the charge density, μ is the magnetic permeability, ϵ is the permittivity of free space, and c is the speed of light. Provided the speeds associated with the fluid motions or with propagating Alfvén waves are much less than the speed of light, the second term on the right side of Equation (4.2) (the displacement current) is small and may be neglected. Equation (4.4) is a consequence of the absence of magnetic monopoles: i.e., all magnetic lines of force must be continuous, and this essential connectivity of magnetic field lines provides many of the problems facing astrophysical theorists. Equation (4.5) implies that electric field lines must terminate on electric charges.

A plasma moving at non-relativistic velocities $\mathbf{u}$ in a magnetic field $\mathbf{B}$ is subject to an induced electric field, $\mathbf{u} \times \mathbf{B}$, which, in the laboratory, gives rise to a current density

$$\mathbf{j} = \sigma(\mathbf{u} \times \mathbf{B}), \qquad (4.6)$$

where σ is the electric conductivity. In the presence of an electric field $\mathbf{E}$ the current density must be proportional to the total field, $\mathbf{E} + \mathbf{u} \times \mathbf{B}$, i.e.

$$\frac{\mathbf{j}}{\sigma} = \mathbf{E} + \mathbf{u} \times \mathbf{B}. \tag{4.7}$$

In many astrophysical applications, such as the highly ionized region below the solar photosphere, the conductivity may be large. In the limit of infinite conductivity, the current density is indeterminate with respect to Equation (4.7) but must satisfy Equation (4.2); that is, it must conform to the curl of the magnetic field, which, because of the non-existence of magnetic monopoles, may be regarded as the primary quality. The electric field must then satisfy Equation (4.7), and, although rapidly neutralized, due to the abundance of charged particles, its spatial variations determine the temporal variation of $\mathbf{B}$, according to Equation (4.3).

In a stationary plasma ($\mathbf{u} = \mathbf{0}$) with high conductivity, $\mathbf{E}$ must be small but exhibit spatial variations consistent with those of $\mathbf{j}$, and the scale-size of this variation, together with the magnitude of σ, determines the 'ohmic' decay of $\mathbf{B}$ through Equation (4.3). Eliminating $\mathbf{E}$ and $\mathbf{j}$ from Equations (4.2), (4.3), and (4.7) yields the *magnetic induction equation*, which governs the changing magnetic field in an astrophysical plasma:

$$\frac{\partial \mathbf{B}}{\partial t} = \nabla \times (\mathbf{u} \times \mathbf{B}) - \nabla \times (\eta \, \nabla \times \mathbf{B}), \tag{4.8}$$

where $\eta = 1/\mu\sigma$ is the *magnetic diffusivity*. Using the zero divergence of $\mathbf{B}$ and assuming that η is constant yields

$$\frac{\partial \mathbf{B}}{\partial t} = \nabla \times (\mathbf{u} \times \mathbf{B}) + \eta \, \nabla^2 \mathbf{B}. \tag{4.9}$$

It is immediately clear that, in a stationary plasma ($\mathbf{u} = \mathbf{0}$), the field must decay in the ohmic decay time $\tau = L^2/\eta$, where L is the scale size of the system. For the Sun's global field Cowling (1976) has estimated a decay time $\tau_\odot \approx 10^{10}$ years, and for sunspots 300 years. However, for highly structured fields, L must correspond to the structure size, and for (say) an active region, τ may be considerably smaller.

If $\eta = 0$, the field $\mathbf{B}$ is completely determined by the plasma motions $\mathbf{u}$, and Equation (4.9) is equivalent to the vorticity equation for an inviscid fluid. Consider the magnetic flux Φ through a material surface S (i.e. a surface that moves with the fluid),

$$\Phi = \int_S \mathbf{B} \cdot \mathbf{dS}. \tag{4.10}$$

If Γ is the material closed curve bounding S, the total rate of change of Φ is

$$\frac{D\Phi}{Dt} = \int_S \frac{\partial \mathbf{B}}{\partial t} \cdot d\mathbf{S} + \oint_\Gamma \mathbf{B} \cdot (\mathbf{u} \times d\boldsymbol{\ell}), \tag{4.12}$$

$$= \int_S \frac{\partial \mathbf{B}}{\partial t} \cdot d\mathbf{S} + \oint_\Gamma (\mathbf{B} \times \mathbf{u}) \cdot d\boldsymbol{\ell}$$

$$= \int_S \left\{ \frac{\partial \mathbf{B}}{\partial t} - \nabla \times (\mathbf{u} \times \mathbf{B}) \right\} \cdot d\mathbf{S}, \tag{4.13}$$

using Stokes' Theorem. If $\eta = 0$ in Equation (4.9), the right hand side of Equation (4.13) is zero. Thus the total flux across any arbitrary surface moving with the fluid remains constant, and the magnetic field lines are said to be *frozen in* to the flow.

In general situations neither $\mathbf{u}$ nor η is zero, and dynamo action, which involves changes of the field relative to the plasma over periods less than the decay time, depends on a delicate interaction between the two terms on the right side of Equation (4.9).

4.2 The fluid equations

The continuity, or mass conservation, equation is

$$\frac{D\rho}{Dt} + \rho \nabla \cdot \mathbf{u} = 0, \tag{4.14}$$

where the total derivative is

$$\frac{D}{Dt} = \frac{\partial}{\partial t} + \mathbf{u} \cdot \nabla,$$

and ρ is the mass density.

The plasma velocity, $\mathbf{u}$, is not a free parameter but is governed by the momentum equation which includes the Lorentz force term, $\mathbf{j} \times \mathbf{B}$, i.e.

$$\rho \frac{D\mathbf{u}}{Dt} = -\nabla p + \mathbf{j} \times \mathbf{B} + \mathbf{F}, \tag{4.15}$$

where p is the plasma pressure and $\mathbf{F}$ is the force term, representing both gravity and the net viscous forces. Assuming a Newtonian fluid with isotropic viscosity, $\mathbf{F}$ may be written as

$$\mathbf{F} = -\rho g(r) \frac{\mathbf{r}}{r} + \rho v \nabla^2 \mathbf{u}, \tag{4.16}$$

where $g(r)$ is the local gravitational acceleration acting in the radial direction and v is the kinematic viscosity.

In a frame of reference rotating with instantaneous angular velocity Ω, at a displacement $\mathbf{r}$ from the rotation axis, Equation (4.15) becomes

$$\rho\frac{\mathbf{Du}}{\mathbf{D}t} = -\nabla p + \mathbf{j} \times \mathbf{B} + \mathbf{F} + \rho\left\{2\mathbf{u} \times \Omega + \mathbf{r} \times \frac{d\Omega}{dt} + \frac{1}{2}\nabla|\Omega \times \mathbf{r}|^2\right\}. \quad (4.17)$$

Since all stars rotate, rapidly when young but more slowly as they age, these modifications cannot be ignored, but, under most circumstances, the latter two terms are small compared with the Coriolis term, $\mathbf{u} \times \Omega$, and may be neglected. We shall see, however, that the Coriolis term plays an important role in understanding stellar activity.

4.3 The equation of state and the energy equation

The first of the remaining equations which determine the constitution of stars is the equation of state, for which the perfect gas law,

$$p = \frac{k\rho T}{m}; \quad = nkT, \quad (4.18)$$

is generally adopted. Here k is Boltzmann's constant, m is the mean particle mass, T is the gas temperature, and n is the number of particles per unit volume.

The second equation describes the flux of energy (heat) through a star and, in its most general form, may be written

$$\rho T\frac{\mathbf{D}s}{\mathbf{D}t} = -L, \quad (4.19)$$

where L is the energy loss function, representing the net effect of all the sinks and sources of energy, and s is the entropy per unit mass of the plasma. There are many more specific forms of this equation, the form most suitable for MHD applications being

$$\frac{\rho^\gamma}{\gamma - 1}\frac{\mathbf{D}}{\mathbf{D}t}\left(\frac{p}{\rho^\gamma}\right) = -\nabla\cdot\mathbf{q} + \kappa_r\nabla^2 T + \frac{j^2}{\sigma} + H, \quad (4.20)$$

where $\mathbf{q}$ is the heat flux due to conduction, κ_r is the coefficient of radiative conductivity, T is the temperature, j^2/σ is the ohmic dissipation (or Joule heating), and H represents the sum of all other heating sources.

4.4 Structured magnetic fields

Practically all the phenomena described in Chapter 3 are directly or indirectly related to magnetic fields, and, since most of these phenomena are highly structured, it must be assumed that the fields themselves are equally structured. Indeed, the field of the magnetic network consists of discrete, intense flux tubes separated by weakly magnetized plasma.

If the plasma velocity is small compared with the sound speed $(\gamma p_0/\rho_0)^{\frac{1}{2}}$, the Alfvén speed $B_0/(\mu\rho_0)^{\frac{1}{2}}$, and the gravitational free-fall speed $(2gl_0)^{\frac{1}{2}}$ (here p_0, ρ_0, and B_0 represent values of the pressure, density, and the magnetic field at some typical level $z = 0$, and l_0 is the vertical scale length), the inertial and viscous terms in Equation (4.15) may be neglected, yielding the equation of magnetohydrostatic balance,

$$0 = -\nabla p + \mathbf{j} \times \mathbf{B} + \mathbf{F}, \tag{4.21}$$

relating the pressure gradient, the Lorentz force, and the gravitational force. This equation must be solved together with Equations (4.2), (4.4), (4.18), and a simplified form of the energy equation. If gravity acts along the negative z-direction and s measures the distance along magnetic field lines inclined at angle θ to this direction, the component of Equation (4.21) in the z-direction is

$$0 = -\frac{dp}{ds} - \rho g \cos \theta, \tag{4.22}$$

or, since $dz = ds \cos \theta$,

$$0 = -\frac{dp}{dz} - \rho g, \tag{4.23}$$

where p and ρ are functions of z along a particular field line. Eliminating ρ between Equations (4.18) and (4.23), and integrating, yields

$$p = p_0 \exp \left\{ - \int_0^z \frac{dz}{\Lambda(z)} \right\}, \tag{4.24}$$

where p_0 is the pressure at $z = 0$ and

$$\Lambda(z) = \frac{kT(z)}{mg} = \frac{p}{\rho g} \tag{4.25}$$

is the *local pressure scale height*. In terms of the density, Equation (4.24) becomes

$$\frac{\rho}{\rho_0} = \frac{T_0}{T} \exp \left\{ - \int_0^z \frac{dz}{\Lambda(z)} \right\}, \tag{4.26}$$

Equation (4.24) shows that the pressure along a given field line decreases with height, the rate of decrease depending on the temperature structure as determined by the energy equation. The magnetic field enters implicitly, since the geometry of the field line depends on the magnetic structure and may influence both the conductive and Joule heating terms in Equation

(4.20). When the temperature increases with height, the density decreases faster than the pressure; but, when the temperature decreases with height, the density may either increase or decrease, depending on the contribution of T to Equation (4.26).

In many applications not all the terms in Equation (4.21) are equally important; for example, if the height of a structure is much less than the pressure scale height, gravity may be neglected compared with the pressure gradient. If, in addition, the ratio β of the gas pressure, p_0, to the magnetic pressure, $B_0^2/2\mu$, is small, i.e.,

$$\beta = \frac{2\mu p_0}{B_0^2} \ll 1, \qquad (4.27)$$

any pressure gradient is dominated by the Lorentz force, and Equation (4.21) reduces to

$$\mathbf{j} \times \mathbf{B} = 0. \qquad (4.28)$$

In this case, the magnetic field is said to be *force free*, and Equation (4.28) requires that either the current $\mathbf{j}$ is parallel to $\mathbf{B}$ (*Beltrami* fields) or that $\mathbf{j}(= \nabla \times \mathbf{B}) = 0$, in which case the field is said to be a *current-free* or *potential* field.

If β is not negligible and the field is strictly vertical, of the form $\mathbf{B} = B(x)\mathbf{k}$ (i.e. gravity is neglected), Equation (4.21) becomes

$$0 = \frac{\partial}{\partial x}\left\{ p + \frac{B^2}{2\mu} \right\},$$

which has the solution

$$p + \frac{B^2}{2\mu} = k(\text{const}), \qquad (4.29)$$

and, for a vertical flux tube within which the field B and pressure p_i are uniform and in balance with the external pressure p_e, Equation (4.29) becomes

$$p_i + \frac{B_0^2}{2\mu} = p_e. \qquad (4.30)$$

This deceptively simple solution conceals the many difficulties encountered when the gravitational term and the stratification of both pressure and density are included in Equation (4.21). For a more comprehensive discussion, the reader is referred to Priest (1982).

4.5 Magnetic buoyancy

Equation (4.30) also applies to a horizontal flux tube of cross-sectional area A with internal and external temperatures and densities, provided that the gravitational free-fall time across the tube is small compared with the Alfvén speed. In this case, Equation (4.30) may be rewritten

$$\rho_i = \frac{T_e \rho_e}{T_i} - \frac{m B_0^2}{2\mu k T_i} \tag{4.31}$$

and, since the net gravitational force per unit volume on such a region (i.e., the buoyancy) is

$$(\rho_e - \rho_i)g = \rho_e \left\{ 1 - \frac{T_e}{T_i} \left(1 - \frac{1}{\beta} \right) \right\} g, \tag{4.32}$$

the buoyancy of the tube is positive provided that

$$\frac{T_i}{T_e} > 1 - \frac{1}{\beta}. \tag{4.33}$$

Although the relationship is more complicated when variations in the vertical direction must be included, this simple analysis demonstrates that horizontal magnetic flux tubes in a gravitationally stratified plasma tend to be buoyant, because, unless the internal temperature is significantly less than the external temperature, the density inside the tube is less than that outside.

4.6 Conclusion

This brief summary of the magnetohydrodynamic and magnetohydrostatic equations is adequate to an understanding of cyclic phenomena. Those interested in exploring further the complex role played by magnetic fields in the phenomena of activity, such as flares, sunspots, spicules, and prominences, are referred to the several treatises on the subject: e.g. Cowling (1976), Parker (1979), or Priest (1982).

References

Cowling, T. G.: 1976, *Magnetohydrodynamics*, Adam Hilger, Bristol.
Parker, E. N.: 1979, *Cosmical Magnetic Fields*, Clarendon Press, Oxford.
Priest, E. R.: 1982, *Solar Magnetohydrodynamics*, Reidel, Dordrecht.

5

The one-dimensional configuration of the cycle

> A point once said to a line
> 'It is a belief of mine
> that it's merely pretension
> to claim a dimension'
>
> *Punch*

5.1 Introduction

The standard concept of the sunspot cycle is of an 11-year variation in the number of sunspots present on the Sun, $N(t)$, at time t. The data from which $N(t)$ must be determined are the daily values of the Zurich sunspot number R_Z (defined in §2.5), but, because only half the Sun is visible at any one time, R_Z is a measure of the number of spots on the visible hemisphere, and it is not possible, even in principle, to determine $N(t)$ at any instant t. For this reason $N(t)$ must be derived from a time average of R_Z over a longer period, such as a Carrington rotation or a calendar year. The variable t is therefore discrete rather than continuous, and the function $N(t)$ is strictly a sequence, N_i, which represents twice the mean value of R_Z during the ith time interval.

The sunspot number R_Z is not the only scalar quantity that exhibits cyclic variations with an 11-year period. Other such quantities include total sunspot area, active region counts, flare counts, the strength of CaII emission, the 10 cm radio flux, the incidence of aurorae, the flux of cosmic rays as measured through certain indicators, and even the widths of terrestrial tree rings. Each of these quantities exhibits a slightly different pattern of variation, and the investigation of the various time series can provide different insights into the nature of the cycle. Such a variety of indicators must, however, raise the question as to what is the fundamental physical variable which generates these variations in secondary phenomena which we call the *solar cycle*.

The traditional determination of N_i using the Zurich sunspot number, R_Z, defined by Equation (2.2), has, however, the considerable advantage of providing a database which extends from the early eighteenth century to the present, although reliability of some of the observations, particularly those made during the seventeenth and early years of the eighteenth century, may be open to question. A plot of the annually averaged value of R_Z against time since 1610, using the best available records as compiled by Waldmeier (1961) and Eddy (1976), is shown in Figure 5.1.

The representation of the cycle as a time series is one-dimensional in the sense that the Sun's behaviour as a whole is referenced to the single parameter t. The butterfly diagram (see Figure 2.1) shows that N has also a more complex representation as a function of two variables, $N(\lambda, t)$, where λ is the latitude of the phenomenon. The two-dimensional representation will be discussed in Chapter 8; here we shall consider properties of the cycle that may be extracted from the one-dimensional representation.

5.2 The Maunder Minimum

Although there is some doubt surrounding the reliability of sunspot number records prior to 1700, all records between 1642 and 1705 indicate that sunspot numbers were greatly reduced in comparison with the period from 1705 to the present, as indicated in Figure 5.1. Late in the last century, both Spörer and Maunder called attention to this 60-year time span during which remarkably few sunspots were reported. Maunder hypothesized that, during this time, now called the 'Maunder Minimum', the normal cyclic appearance of sunspots sank into an intermission, and he queried whether the modern sunspot cycle is really a universal property of the Sun.

The authenticity of this intermission was questioned on the grounds that, since it occurred so soon after the discovery of sunspots, it may have reflected the lack of systematic observations rather than an absence of spots. A detailed investigation by Eddy (1976), who correlated existing sunspot data with auroral data, carbon-14 tree-ring data, and even climatology data, has, however, demonstrated the probable reality of the Maunder Minimum. Weiss (1993) maintains that contemporary astronomers were in little doubt that there was a real shortage of spots during the period and that they reported excitedly on the occasional spots when they appeared. Weiss and Weiss (1979) note that even the poet, Andrew Marvell, in his satirical poem *The Last Instructions to a Painter*, composed in 1667, indicates not only a common awareness of the phenomenom of sunspots:

So his bold Tube, Man to the Sun apply'd

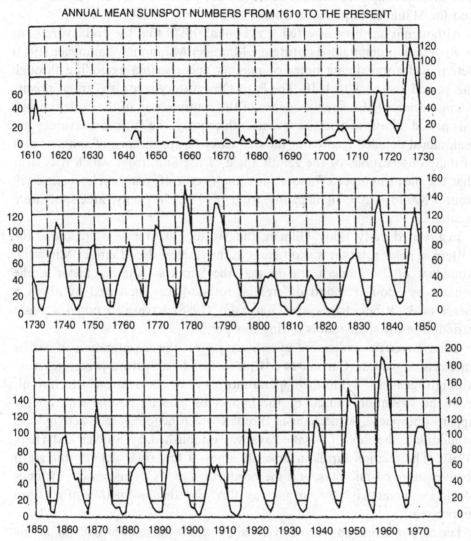

Fig. 5.1 The annual mean of the Zurich sunspot number, R_Z, since 1610, from Waldmeier (1961) and Eddy (1976).

> And Spots unknown to the bright Star descry'd,

but also of their apparent disappearance:

> Through Optik Trunk the Planet seem'd to hear,
> And hurls them off, e'er since, in his career.

Lalande and Herschel were aware of this intermission period, but, apparently, it was overshadowed by Schwabe's discovery of the cycle in 1843 (§2.3) and subsequently forgotten. It remained for Spörer to rediscover the effect and for Maunder to publicise Spörer's work.

Although it has been asserted (e.g. Foukal, 1990) that the cyclic variations of R_Z were not maintained during the Maunder Minimum, Waldmeier (1941) determined approximate times of minima and maxima extending through this period and back to 1610. The Paris Observatory records (Parker, private communication) also show evidence of the continuation of the cycle through this period. Since early in the seventeenth century, the Paris Observatory has maintained detailed records of the frequency of appearance of spots as part of daily observations of the zenith angle of the Sun, from which it is clear that the mean number of spots emerging over a 10-year (cycle) period fell from ~ 800 to ~ 200. Significantly, from 1690 to 1700, only three spots were recorded.

In a related study, Eddy, Gilman, and Trotter (1976) have claimed that the pattern of solar rotation was different at the start of the Maunder Minimum from that of the modern era, in that the Sun was rotating faster at the equator by about 3% and differential rotation was increased by about a factor of 3. Ribes, Ribes, and Barthalot (1988) have also noted a slight variation in the solar radius during this period.

Other terrestrial indicators of activity permit the investigation of earlier intermission periods. Carbon-14 (^{14}C) is formed by a neutron-proton reaction with nitrogen-14 (^{14}N) in the upper atmosphere by the neutrons which result from the spallation caused by the impact of galactic cosmic rays on the upper atmosphere. A mechanism through which heliomagnetic activity acts to modulate the flux of cosmic rays was established by Stuiver and Quay (1980). The relationship is inverse: at times of low solar activity the earth receives more cosmic rays, and the production of ^{14}C is increased; at times of high solar activity the cosmic ray flux and the production of ^{14}C are suppressed.

The carbon atoms so formed in the upper atmosphere eventually find their way into plants, where the atoms are assimilated and preserved, long-term changes in the atmospheric CO_2 being recorded in ^{14}C tree-ring data, in which the ratio of ^{14}C to ^{12}C correlates well with the envelope of the Zurich sunspot-number count since 1700. Using this and other results, Eddy has demonstrated the probable occurrence of other 'intermission' periods during the fifteenth (the 'Spörer' minimum) and fourteenth (the 'Wolf' minimum) centuries, and a 'medieval maximum' period during the twelfth century.

Variations in the tree-ring spacings found in certain areas of the south-western United States are also thought to be caused by solar-wind induced shifts in the boundaries of atmospheric jet streams, which can affect the local

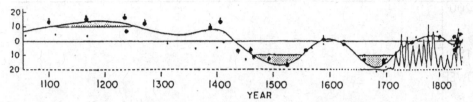

Fig. 5.2 Variations in the percentage of ^{14}C with respect to ^{12}C since 1050 AD are shown together with the annual mean of R_Z between 1700 and 1850. From left to right the shaded regions represent (i) the Medieval Maximum, (ii) the Spörer Minimum, and (iii) the Maunder Minimum (from Eddy 1976).

pattern of rainfall. The tree-ring spacings have also been used to construct proxies for cyclic activity prior to the seventeenth century. Figure 5.2 shows the solar activity amplitude envelope extending back to the year 1000, as inferred from the combination of sunspot observations, tree-ring data, and incidence of northern-hemisphere aurorae. The correspondence between the ^{14}C curve and the envelope of the sunspot number plot since 1700 supports the use of this curve as a proxy for earlier periods of both normal and abnormal sunspot activity. The presence of a 200-year periodicity in the tree-ring data also has been recently claimed by Runcorn (private communication).

Eddy also noted that the time of the Maunder Minimum was one of unusual cold in Europe. The Paris archives show that the average temperature was reduced by $1\frac{1}{2}$% during this period, which has become known as the 'Little Ice Age'. The period 1690–1700, when only three spots were observed, was indeed one of extreme cold, which gave rise to crop failures and famine.

The coincidence of the Little Ice Age with the Maunder Minimum may be just that, a coincidence. However, recent results summarized by Foukal (1992) indicate the possibility of a closer connection (see § 5.6).

5.3 The cycle since 1700

The plot of the annually averaged value of R_Z since 1705, shown as part of Figure 5.1, makes it clear that, although all cycles have periods close to 11.2 years, no two cycles are identical. Indeed, they vary considerably, not only in amplitude but also in period and shape. The stronger cycles, e.g. 19 and 21, tend to rise more rapidly to maximum and have a shorter overall period, while weaker cycles tend to be flat-topped, a precise maximum being difficult to determine. These aperiodic features of the sunspot cycle curves are suggestive of a non-linear system on the verge of *chaos*, a possibility explored in Chapter 13. There is some suggestion of a longer periodicity

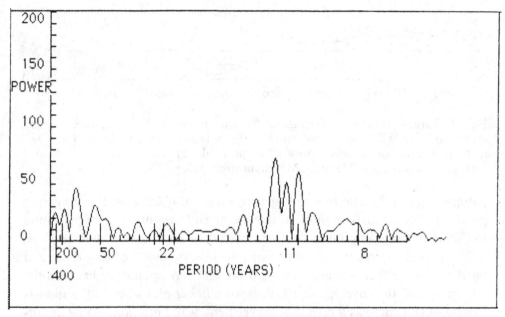

Fig. 5.3 The power spectrum of the annually averaged values of R_Z since 1705 is plotted against frequency and the appropriate periodicities (in years) are indicated on the abscissa.

in the amplitude modulations. Wolf (1862), for example, noted an apparent periodicity of ~ 80 years, which seems to be continued in the twentieth-century observations by Brunner–Hagger and Liepert (1941), but requires a longer time span for confirmation.

The usual, and some unusual, statistical techniques have been applied for the analysis of the time-series N_i. Figure 5.3 shows a power spectrum, obtained from the Fourier transform of the annually averaged sunspot number time-series since 1705, as plotted in Figure 5.1. This transform is defined by

$$P(v) = \int_0^T |N(t)e^{ivt}|^2 dt.$$

The variability of the cycle is well illustrated by the complexity of this power spectrum. Most power is concentrated in a range of frequencies corresponding to periods from 8 to 16 years; three peaks, each of comparable strength, correspond to periods of 10, 11, and 12 years. Also seen in this spectrum are smaller peaks, corresponding to 22, 40, 80, and even 200-year periods. Since the total time interval occupied by these data is only 285 years, the last of these periods may be spurious, arising as a result of *aliasing effects*, whereby periodicities related to the finite length of the data sample may appear in the power spectra. Inclusion of the Maunder Minimum period in the data only distributes the power more evenly over the frequency range, and the power at longer periods (100–200 years) becomes comparable with that at the 8–16 year range.

5.4 The alternating cycle

An ingenious modification of the sunspot number plot has been proposed by Bracewell (1985, 1986). Arguing that, since consecutive sunspot cycles exhibit opposite polarity signatures (both in the polarities of sunspot pairs and in the polar fields), he proposes that sunspot counts should be signed, with spots of successive cycles having alternate signs.

The minima of the unsigned plot tend to be sharp as compared with the maxima, but it is evident from the butterfly diagram (Figure 2.2) that these minima actually represent the mixture of two overlapping cycles, which are distinguished by latitude in the diagram– one waxing while the other wanes – rather than a lull between one cycle and the next. Thus, when the 'signed' sunspot number plot is constructed, the point at which the 'signed' sunspot number should cross the axis can only be determined with the aid of the butterfly diagram: by subtracting the number of spots associated with the 'negative' cycle from those associated with the 'positive'. The plot of signed sunspot numbers is compared with the standard plot since 1705 in Figure 5.4. The association of pairs of strong (e.g. 18 and 19) and weak cycles (e.g. 10 and 11) now appears natural, and although the sample is insufficient to permit any firm conclusion, there are suggestions of an irregular beat pattern extending over 5–6 cycles.

A significant change is seen in the power spectrum when it is calculated for the alternating cycle (see Figure 5.5). The sign alternation embraces the alternate pairings of strong and weak cycles and this, of course, eliminates the 10–12 year periods from the power spectrum. Virtually all power is now concentrated in the frequency range corresponding to 18–25 years, with the peak at 22 years clearly the dominant one. No power is found at frequencies corresponding to 200 years, nor, perhaps surprisingly, at 80–100 years. This dramatic simplification of the power spectrum calculated for the alternating diagram supports the concept that sunspot numbers, as well as flare counts, CaII emission measures, etc., are merely convenient scalar indicators of a cycle which is essentially a signed magnetic cycle, with a fundamental periodicity of 22 rather than 11 years.

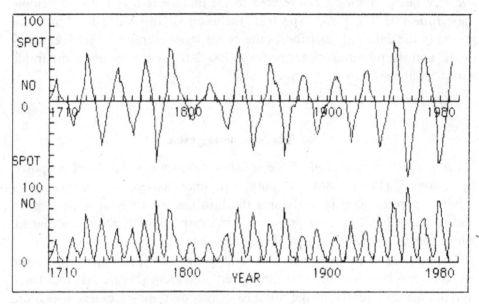

Fig. 5.4 The lower panel shows the annually averaged values of R_Z between 1705 and 1986, while, in the upper panel, negative values are assigned to R_Z in alternate cycles, as suggested by Bracewell (1986).

5.5 The north–south asymmetry

Although the cyclic behaviour of the two hemispheres is obviously related by the Hale–Nicholson polarity law, the time variations of the various activity parameters exhibit varying degrees of asymmetric behaviour. The reversals of the polar fields may show a time difference of as much as two years (e.g. in Cycle 19 the reversal of the southern polar field followed two years after that of the northern field). A similar lag may occur in the emergence of the first new-cycle spots in opposite hemispheres (e.g. Cycle 20).

Asymmetries in the amplitudes of various phenomena are also reported. Bell (1962) found a long-term asymmetry in the sunspot area data for Cycles 8 through 18. Roy (1977) studied the north–south distribution of flares for a period of two cycles and found an asymmetry in favour of the northern hemisphere that increased with the importance of the solar event. Reid (1968) also reported asymmetry in favour of the northern hemisphere for the period 1958–1965. Howard (1974) found that the gross magnetic flux in the northern hemisphere exceeded that in the south by $\sim 7\%$ over the period 1967–1973. Verma (1987) found that the asymmetries in major flares, white-light flares, and type II radio bursts favoured the northern hemisphere in Cycles 19 and 20, while asymmetries in type II bursts, white-light flares,

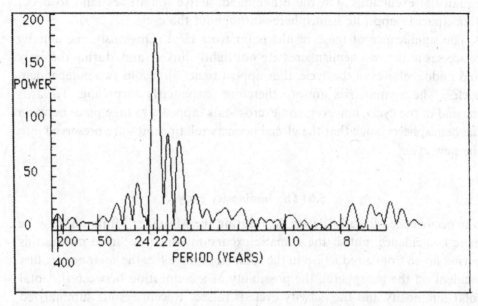

Fig. 5.5 The power spectrum of Bracewell's signed annually averaged values of R_Z since 1705 is plotted against frequency and the appropriate periodicities (in years) are indicated on the abscissa.

gamma ray bursts, hard X-ray bursts, and coronal mass ejections favoured the southern hemisphere during Cycle 21.

Verma (1992) has also calculated the N–S asymmetry indices A, defined by

$$A = \frac{A_n - A_s}{A_n + A_s},$$

where A_n and A_s are the appropriate annual measures for each of seven particular activity indicators in the northern and southern hemispheres for the period embracing Cycles 8–22. He finds that the N–S asymmetry may have a period of 12 solar cycles, a period uncomfortably close to the total sample size, but he bravely predicts that the N–S asymmetry may be southern-dominated in Cycles 22, 23, and 24 and then revert to northern domination in Cycle 25.

K. Harvey (private communication) has found little correlation between the emergence of active regions at similar latitudes, but in opposite hemispheres, during the first nine years of the cycle, and notes an increasing trans-equatorial correlation as sunspot minimum approaches. It is during this phase that the major neutral lines which encircle the higher latitude regions during the former period tend to deform and take part in trans-equatorial excursions. On the other hand, active longitudes tend to occur 180° apart in opposite hemispheres throughout the cycle.

The significance of these results is far from clear. Obviously, the activity processes in the two hemispheres are not tightly linked, and, during the early and middle phases of the cycle, they appear to act almost as two independent cycles. The asymmetries are not therefore particularly surprising. Towards the end of the cycle, however, some 'cross-talk' appears to take place between the hemispheres, such that the global polarity relationships are preserved into the new cycle.

5.6 The luminosity cycle

The occurrence of the Little Ice Age during the Maunder Minimum may be mere coincidence, but, if the climatic excursion which occurred during this period arose from a reduction in the solar irradiance (i.e. the solar energy flux incident on the geosphere), the possibility of a connection between the total solar luminosity and the activity cycle is raised. Recent results summarized by Foukal (1992) explore this possibility.

The best data on solar irradiance variations have been obtained by the ERB radiometer on the Nimbus 7 satellite (Hickey *et al.* 1988) and by the ACRIM radiometer on the SMM spacecraft, beginning in early 1980 and continuing to late 1989, with a short interruption between late 1983 and early 1984 (Willson and Hudson 1991). These observations show (i) a high frequency variation on time scales of about 10 days, with peak-to-peak amplitudes reaching about 0.3%, and (ii) a gradual decline of the mean irradiance between 1980 and 1986, followed by its recovery through 1990. The decline observed by both the ERB and ACRIM between 1980 and 1986 is less than 0.1%.

It is easy to identify the large dips in the total irradiance with the passage of sunspot groups across the disk. These dips can be accounted for qualitatively by a simple model, which assumes that the energy deficit of the sunspots is blocked and stored in a small increase in the thermal and potential energy of the convecting layers outside the sunspot over the time-scale of the variations.

If the effect of sunspot blocking is subtracted from the measured irradiance, there are significant residuals which Foukal and Lean (1986) have shown to be related to the passage of photospheric faculae across the disk. Their brightness temperature near the limb exceeds that of the photosphere, and, using an empirical model which incorporates both the heat-blocking effect of sunspots and the excess brightness of faculae (including those associated with magnetic networks outside active regions), it is possible to match not only the 6–9 month variations due to magnetic activity but also the slow decrease in the irradiance from 1980 to 1986. Since a steady rising trend has been in progress since 1986, it would appear that, for this period at least, a luminosity fluctuation accompanies the activity cycle and is in phase with it. Although of small amplitude (< 0.1%) this fluctuation is significant in relation to stellar activity cycles (see §7.12), because it implies that the excess emission from faculae slightly outweighs the blocking effect of sunspots over the medium-to-long term.

This result, together with the concurrence of the Maunder Minimum and the Little Ice Age, has fuelled speculation that sunspot activity may be related to total energy output and that, in counter-intuitive fashion, periods of greater sunspot activity correspond to increased energy output. If so, an understanding of the nature of such 'grand minimum periods' and, more particularly, the ability to predict such occurrences, could have considerable climatological significance.

5.7 The 'varves' hypothesis

While the tree-ring spacings and ^{14}C data have permitted the extension of the sunspot 'proxy' data-base back to 1100, and perhaps a few centuries earlier, the total data-base potentially available is still far too short to apply more sophisticated analyses and tests for chaotic behaviour, such as those discussed in Chapter 13. Thus it was with some excitement that the possibility of obtaining proxy data from the pre-Cambrian era was seized upon during the 1980s.

While prospecting in the Flinders Ranges of South Australia in the 1970s, George Williams, a geologist with the BHP Company of Australia and an amateur astronomer, came upon some pre-Cambrian sandstone rocks which displayed a layering pattern exhibiting a periodicity of 12 ± 3 layers in the relative thicknesses of the layers. Williams (1981), and Williams and Sonnet (1985) proposed that these were annual layerings, or 'varves', reflecting annual climatic changes whose strengths were related to the solar activity. These patterns were not unlike the tree-ring spacings referred to above. The periodicity of ~ 12 years appeared to indicate that a solar cycle signal (albeit of slightly longer duration) was reflected in these pre-Cambrian records, and that therefore, by studying the variations between successive batches of layers, the database of the quasi-periodic cycle could be extended dramatically from a few hundred years in the present period to many thousands of years in the pre-Cambrian.

Many solar physicists seized on this possibility, ran batteries of statistical tests, and announced the existence of an 'Elatina' period of 300 years and a longer period of 350 years. Excitement ran high, and soon a whole new industry of 'sandstone' astrophysics was in progress. Beat frequencies corresponding to competing 11- and 12-year cycles were proposed, and, in a careful time-series analysis, it was shown (Bracewell 1986) that a remarkably precise timing mechanism must be involved. The old idea of the presence of a magnetic oscillator lying below the convection zone was resurrected (see Chapter 7) and the possibility that such ancient data could yield long-term predictions of solar activity during the present epoch advanced.

In the midst of this flurry of activity, scepticism was growing about the interpretation of these data as a proxy for the solar cycle. The timing was *too* precise; no periods of extended minima were found in the pre-Cambrian data, so that the solar interpretation cast new doubts on the existence of the Maunder Minimum and other minima as well. Further, the indication from these data that the period of the ancient solar cycle was $\sim 8\%$ longer than at present contradicted the predictions based on stellar models that it should have been $\sim 3 - 10\%$ shorter (Noyes *et al.* 1984). Finally, the discovery of another series of ancient layerings in sandstone (Preiss 1987), which contained still longer periods with 14–15 layers per cycle, forced many to reassess the varve interpretation.

The end followed swiftly: first Sonnet pointed out, and soon Williams agreed (Sonnet and Williams 1987), that the layers, with their precise timings and beat frequencies, were more typical of (monthly) lunar variations in tidal deposits. Before the solar sandstone physics community could react, Williams (1988) presented a barrage of arguments which were now just as persuasive in favour of the tidal interpretation as had been his original arguments for the solar-cycle connection: the sandstone layers were not 'varves' at all and could tell us nothing about the pre-historical sunspot cycle. The new industry, like so many others which seemed like a good idea at first, collapsed forthwith.

One possibility for dramatically extending the data base of cyclic behaviour was thus closed off. On the other hand, a very different approach, sometimes referred to as the *solar–stellar connection*, had begun in the 1960s and, during the 1980s, began to yield exciting results which have greatly extended the data concerning the long-term changes in cyclic activity in stars. These results are discussed in Chapter 7.

5.8 Conclusion

Although it provides only a partial description of solar cyclic behaviour (see
Chapter 8 for the two-dimensional cycle and Chapter 13 for a discussion
of chaos and the cycle), the one-dimensional account contains important
data which define the parameters of the cycle. The power spectra calculated
on the basis of Bracewell's inversion of the odd sunspot cycles strongly
suggests that the 22-year magnetic cycle is the fundamental cycle (since the
peaks in its power spectrum are more cleanly defined). From this it may be
inferred that some deep-seated magnetic configuration is the fundamental
physical variable and that the surface phenomena, such as the variations in
the sunspot number, or sunspot area, or flare count, or CaII emission, or
any of the other activity phenomena are only the surface signatures of this
fundamental magnetic cycle.

Some simple phenomenological models which account for some, but not
all, of these data are now considered.

References

Bell, B.: 1962, *Smithsonian Contr. Astrophys.*, **5**, 187.
Bracewell, R. N.: 1985, *Australian J. Phys.*, **38**, 1009.
Bracewell, R. N.: 1986, *Nature*, **323**, 516.
Bruner–Hagger, W., and Liepert, A.: 1941, *Astr. Mitt. Zurich*, **140**, 556.
Eddy, J. A.: 1976, *Science*, **192**, 1189.
Eddy, J. A., Gilman, P. A., and Trotter, D. E.: 1976, *Solar Phys.*, **46**, 3.
Foukal, P.: 1990, *Sci. Am.*, **262**, 34.
Foukal, P.: 1992, in *Proceedings of the National Solar Observatory/Sacramento Peak
      12th Summer Workshop*, ed. K. L. Harvey, San Francisco, California, 439.
Foukal, P., and Lean, J.: 1986, *Ap. J.*, **302**, 826.
Hickey, J., Alton, B., Kyle, H., and Hoyt, D.: 1988, *Space Sci. Rev.*, **48**, 321.
Howard, R.: 1974, *Solar Phys.*, **38**, 59.
Noyes, R. W., Weiss, N. O., and Vaughan, A. H.: 1984, *Astrophys. J.*, **287**, 769.
Preiss, W. V.: 1987, *The Adelaide Geosyncline*, Bull. **53**, S. Aust. Dept. Mines and
      Energy, Adelaide.
Reid, J. H.: 1968, *Solar Phys.*, **5**, 207.
Ribes, E., Ribes, J-C., and Barthalot, R.: 1988, *Nature*, **332**, 689.
Roy, J. R.: 1977, *Solar Phys.*, **52**, 53.
Sonnet, C. P., and Williams, G. E.: 1987, *Solar Phys.*, **110**, 397.
Stuiver, M., and Quay, P. D.: 1980, *Science*, **207**, 11.
Verma, V. K.: 1987, *Solar Phys.*, **114**, 185.
Verma, V. K.: 1992, in *Proceedings of the National Solar Observatory/Sacramento
      Peak 12th Summer Workshop*, ed. K. L. Harvey, San Francisco, California, 429.
Waldmeier, M.: 1941, *Astr. Mitt. Zürich*, **14**, 551.
Waldmeier, M.: 1961, *The Sunspot-Activity in the Years 1610–1960*, Schulthess and
      Co., Zurich.
Weiss, J. E., and Weiss, N. O.: 1979, *Q. J. Roy. Astron. Soc.*, **20**, 115.

Weiss, N. O.: 1993, in *NATO Workshop Proc.* (to appear).
Williams, G. E.: 1981, *Nature*, **291**, 624.
Williams, G. E.: 1988, *Climate Change*, **13**, 117.
Williams, G. E, and Sonnet, C. P.: 1985, *Nature*, **318**, 523.
Willson, R., and Hudson, H.: 1991, in *Proceedings of the SOLERS 22 Workshop*,
    June 3–7, 1991, Boulder, Colorado, ed. R.F. Donnelly.
Wolf, R.: 1862, *Astr. Mitt. Zürich*, No 14.

6

Heuristic models of the solar activity cycle

round up the usual suspects
Casablanca

6.1 Introduction

Although the solar cycle was identified as a sunspot number fluctuation in 1850, and as a magnetic oscillation in 1923, little progress was made in developing theories which might explain the somewhat diverse phenomena associated with it until the latter half of this century. Since 1950, however, several ingenious, and essentially heuristic, models have been proposed, and some have subsequently been supported by more detailed mathematical analyses. Each of these has offered some insight into possible cyclic processes, but none has provided an account consistent with all the available data, or with our current understanding of the physical processes operating within the Sun. Nevertheless, in order to set the stage for later, more mathematical discussions, it may be helpful to review these models briefly in order to see where they have succeeded and to understand how they have failed.

These models may be classified as (i) relaxation models (e.g. Babcock 1961), (ii) forced oscillator models (e.g. Bracewell 1988), or (iii) dynamo wave models (e.g. Parker 1955, Krause and Rädler 1980). Both (i) and (iii) may be regarded as particular examples of the formal mathematical discipline known as dynamo theory, which will be discussed in more detail in Chapter 11. The models to be discussed here have all arisen from attempts to understand the phenomena of the solar cycle and are part of the background against which more recent observational data should be considered.

6.2 The Babcock model

The best known relaxation model is that proposed by Horace Babcock (1961), and, because of its historic importance, it will be discussed in some detail.

In this model, illustrated in Figure 6.1, an initially global poloidal magnetic field (stage 1) is wound by differential rotation into a spiral field with a strong toroidal component (stage 2). As the winding is increased, Ω-loops or kinks form in the toroid and float to the surface, where they emerge as active regions (stage 3), give rise to sunspot groups and other manifestations of activity, and then decay.

An essential feature of relaxation models is that, as a result of this decay, a new global poloidal field of opposite polarity is established at or before the next solar minimum, prior to the start of the winding process of the new cycle. The development of this field marks the beginning of the cycle, and models starting with this global field can account for the temporal synchrony and subsequent active-region polarity relationships between the hemispheres.

Babcock's original model postulated that, as a bipolar active region decayed, the flux loop connecting leader and follower polarity regions expanded into the corona, and the foot-points moved apart. It is well known that, on average, the magnetic axis of a new active region (from follower to leader) is initially tilted equatorward by $\sim 12°$. Babcock argued that, if this tilt were maintained as the foot-points continued to move apart, the follower flux should tend polewards, cancelling with the existing polar field, while the leader flux would tend equatorwards, where cancellation should take place with leader flux from the opposite hemisphere. A trans-equatorial poloidal loop should then form, connecting follower flux from one hemisphere with that of the other (Figure 6.1 stage 4). The accumulation of such loops during the cycle gives rise to the cancellation of the old and the development of the new polar poloidal field.

This picture explained, in disarmingly simple terms, those features of the cycle which had for so long puzzled solar physicists: the reversed magnetic polarities of sunspot pairs in opposite hemispheres and in consecutive cycles, the initial equatorward tilt of the magnetic axis, the reversal of the polar fields and their intimate involvement with the cycle, and, to a lesser extent, the butterfly diagram.

It was shown that the magnetic tension, and thus the buoyancy of the toroidal field, was greatest at mid-latitudes, so that the first active regions of a new cycle emerged there. It was, however, far from clear why the subsequent regions emerged along the equatorward branch of the toroid but

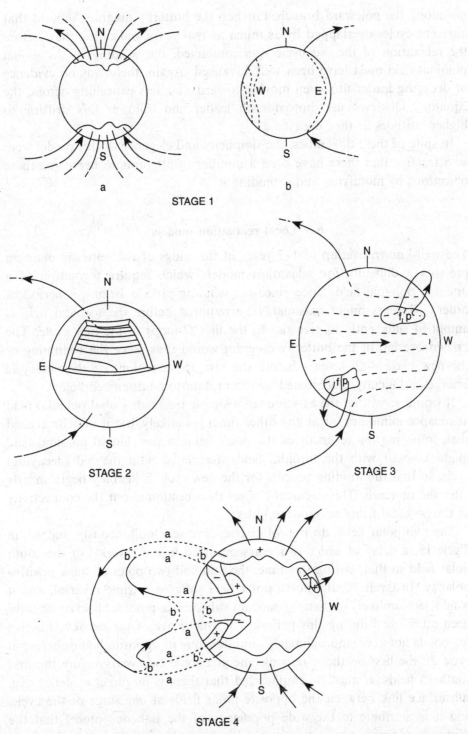

Fig. 6.1 Four stages in the evolution of the global solar magnetic fields according to the Babcock model are illustrated by sketches adapted from Babcock (1961). In the diagram representing stage 3, the letters 'p' and 'f' refer to the foot-points of the field associated with a preceding or leader spot and that associated with a follower spot, respectively.

not along the poleward branch. Further, the butterfly diagram showed that adjacent cycles overlapped by as much as two years, indicating that, before the relaxation of the old cycle was completed, the winding of the global poloidal field must have been well advanced. Again, there was no evidence of decaying leader flux even moving towards, far less cancelling across, the equator. Observations showed both leader and follower flux drifting to higher latitudes as they decayed.

In spite of these difficulties, the simplicity and elegance of this model were so attractive that there have been a number of attempts to overcome these objections by modifying and extending it.

6.3 Local relaxation models

The well-known overlap of 1–2 years of the wings of the butterfly diagram presents a difficulty for relaxation models which require formation of a uniform poloidal field, since, once the winding process begins, a period of order 2–3 years might reasonably be required before the toroidal field is amplified sufficiently to give rise to the first Ω-loops of the new cycle. The known overlap of the butterfly diagram would thus place the beginning of the new cycle $\sim$ 3–4 years before the last active regions of the old cycle emerge, a difficulty exacerbated by recent data to be discussed below.

It would seem that the existence of a simple (relaxed) global poloidal field at sunspot minimum (or at any other time) is unlikely, but it may be argued that, following the reversals of the polar fields, a new global poloidal field might co-exist with the toroidal fields associated with the old (decaying) cycle, so that the winding process for the new cycle fields may begin shortly after the reversals. This argument raises the question about the connectivity of the poloidal fields at opposite poles.

The two polar fields do not, of course, reverse simultaneously; indeed, in Cycle 19, a delay of order two years occurred in the reversal of the south polar field so that, during this time, the Sun had two poles of weak positive polarity. In Cycle 20, the north polar field suffered a triple reversal, and it would seem unlikely that any significant sub-surface poloidal field could have been established during this period. The emergence of the new-cycle active regions is not even approximately simultaneous in opposite hemispheres; in cycle 20 the first northern regions emerged some two years before the first southern fields. It must be emphasized that there is *no* direct evidence of a subsurface link between the opposite polar fields at *any* stage of the cycle, and it is a tribute to the wide popularity of the Babcock model that the existence of such a connection is widely accepted.

In these models, the decaying fields may relax on a local or piecemeal basis. For example, some high-latitude phenomena might arise from the decay of the active regions emerging during the growth phase of the current (old) cycle, while the lower-latitude phenomena of the new cycle result from the post-maximum active regions of the old cycle. Nevertheless, it is hard to see why the consequences of the decay of one set of active regions should be so different from those of another, and, in particular, why the high-latitude phenomena are so much weaker than those at lower latitudes. (This is an objection which can be levelled at many models, including Babcock's.)

Such models have not, as yet, been seriously put forward in the literature, and, in the meantime, it is more profitable to explore other possibilities.

6.4 Preferential poleward migration

Giovanelli (1985) argued that, if the initial equatorward tilt is maintained during the subsequent decay of the region, the follower flux should reach the polar regions ahead of the leader flux and begin the cancellation process. Replacement of the cancelled polar flux by the like-polarity leader flux, which had also drifted polewards behind the follower flux, might restore the *status quo*, but Giovanelli argued that this leader flux might well be cancelled by the next wave of follower flux, resulting in the failure of the restoration of the original field.

The continuation of this process during the cycle should result in the oscillation of the polar field but not its sustained reversal. For this reason Giovanelli was also forced to postulate the equatorward drift and trans-equatorial cancellation of some leader flux so that, eventually, the net excess of follower flux would effect the reversal.

6.5 The flux transport model

A more sophisticated numerical version of this general approach was developed first by Leighton (1964) and has been subsequently extended by Sheeley and his co-workers (e.g. DeVore, Sheeley and Boris, 1984, Sheeley, Nash and Wang, 1987, DeVore and Sheeley, 1987, Wang, Nash and Sheeley, 1989). The model is based on the diffusive decay of active regions under the influence of random supergranule velocity fields, assisted by large-scale meridional flows and differential rotation. Leighton described the process as a 'random walk', defined only the 'number surface density, n, of points at which lines of force enter the Sun', and derived his fundamental equation in terms of n.

DeVore *et al.* (1984) rewrote Leighton's equation as the 'flux transport equation' in terms of the radial component, B_r, of the large-scale fields, viz.

$$\frac{\partial B_r}{\partial t} + \nabla \cdot (B_r \mathbf{v}_s) = \kappa \nabla_s^2 B_r. \qquad (6.1)$$

Here $\mathbf{v}_s$ is the surface component of the large-scale velocity field, κ is the kinematic diffusivity due to the supergranule motions, and ∇_s^2 is the two-dimensional Laplacian operator in the plane of the surface.

Sheeley and his colleagues generated numerical solutions of this equation, taking as initial conditions a two-dimensional grid of values derived from observation and integrating forwards in time in order to obtain simulated distributions of the fields, using appropriate values of the diffusivity and models for the large-scale velocity fields.

This model, hereafter cited as the 'flux transport model', has achieved considerable success in reproducing some of the patterns of evolution of the large-scale field. Further, the correlation-tracking of granules indicates that the transport of magnetic flux is well correlated with horizontal photospheric flows.

The model is, however, susceptible to the criticism that it emphasizes only surface phenomena and neglects the essential physics of the interactions below the surface. Although it requires that surface flux be transported across distances comparable with the solar radius, the model considers only the radial component of the magnetic field, regarding it as a *scalar* quantity and treating the surface field elements as if they were, in effect, 'corks' carried along by an eddying stream.

However, the bipolar magnetic fields which appear at the solar surface in active regions are assumed to arise from the emergence of a loop or stitch of flux, sometimes characterized as an 'Ω-loop', in a postulated sub-surface toroid anchored at the base of the convection zone. The 'cork' model ignores the essential connectivity of magnetic field lines required by Maxwell's law

$$\nabla \cdot B = 0 \qquad (6.2)$$

and the associated magnetic tensions. Indeed, while the surface fields remain connected to this toroid, a more appropriate analogy is with a marine 'buoy' anchored to the bottom of a tidal flow by a cord or chain. When the buoyant surface elements of these fields are displaced through any significant distance by the drag of a transverse flow, such as the meridional circulation, the induced curvature of the field lines must give rise to a net transverse component of the magnetic tension which, acting on an element of the flux tube, may balance the drag force of the flow.

The magnitude of this effect may be estimated by considering an element of the flux tube of radius r, length δh, and field B. Neglecting diffusion and considering only the effect of a transverse flow v, the drag force δF is

$$\delta F = C\rho v^2 r\delta h, \tag{6.3}$$

where ρ is the plasma density and C is the drag coefficient of order unity. If, as a result of this drag, this element is inclined to the vertical at angle θ at its upper end and at $\theta + \delta\theta$ at the lower, the component of the magnetic tension opposing the drag δT_m is

$$\delta T_m = \frac{1}{4}r^2 B^2 \cos\theta\delta\theta. \tag{6.4}$$

Thus the drag force is balanced when

$$\delta\theta = \frac{4C\rho v^2 \delta h}{rB^2 \cos\theta}. \tag{6.5}$$

For a flux tube with field strength of order 10^3 G and of radius ~ 1000 km embedded in a plasma of mean density 4×10^{-4} gm cm^{-3} with a transverse flow of 10 m s^{-1}, and taking C as unity, a distortion $\delta\theta$ of only 0.016 in an element of length 10^4 km would be sufficient to permit the net magnetic tension to balance the drag force. It is therefore hard to see how meridional flows of order 10 m s^{-1} could significantly deform the subsurface flux tubes associated with the surface fields, far less transport them to the polar regions in order to contribute to the polar reversals. Further, even if significant deformation occurs, the net transverse component of the magnetic tension should increase until it balances the drag force of the flow, just as the anchor chain of a floating buoy is deformed.

6.6 The formation and escape of U-loops

The observed surface fields nevertheless *appear* to undergo significant displacements, and it may be that such deformations lead to reconnections between adjacent elements of the fields which release the fields from the constraints of the deep magnetic toroids. Since field-line reconnections in the corona were observed by Skylab and are also seen in chromospheric Hα observations, there is no reason why they should not also occur below the photosphere.

An example of reconnection, leading to the formation of a U-loop, has been proposed by Wilson, McIntosh, and Snodgrass (1990) and is illustrated in Figure 6.2(upper). The process is consistent with one proposed by Spruit *et al.* (1987), which they call the 'repair' of the Ω-loop. Spruit and his

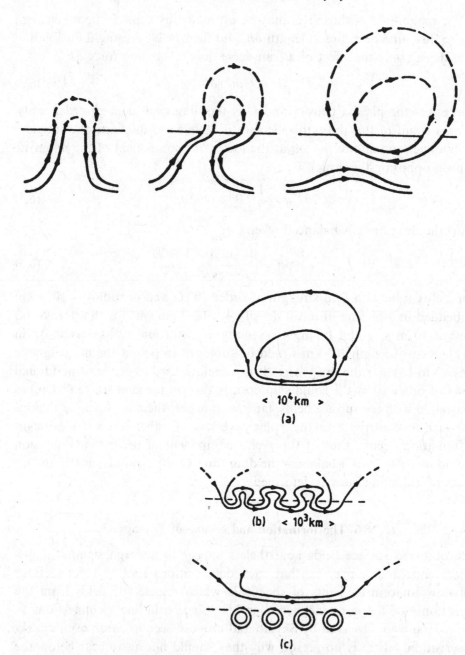

Fig. 6.2 Upper: the emergence of an Ω-loop and the formation of a U-loop by reconnection. Lower: the escape of a U-loop through the surface. (a) The U-loop is trapped at the surface. (b) Small Ω-loops penetrate across the surface and expand as plasma drains down the 'U'. (c) Reconnections release the U-loop above the surface, leaving behind small 'doughnut'-loops which decay rapidly (from Wilson *et al.* 1990).

colleagues argue that, once the loop is freed of the constraints of the deep magnetic toroids, buoyancy would rapidly bring the loop to the surface, where it would be susceptible to the large-scale surface motions. The further role of these U-loops is, however, far from clear.

Spruit *et al.* have also suggested that the shallow horizontal flux at the base of the 'U' would be quickly removed by what they characterize as the 'sea-serpent' process (see Figure 6.2(lower)), as a result of which these fields would be unable to take part in the polar field reversals. Parker (1984), on the other hand, believes that purely toroidal fields may be unable to escape through the surface because of the difficulty of disengaging the field from the highly conducting gas in which it is embedded. While this process may permit the retention of these fields long enough to contribute to the reversals, it raises more serious problems for stars if they are unable to divest themselves of excess toroidal fields. Further, while the follower flux might reach the pole ahead of the leader flux and cancel with the existing field, the leader flux, necessarily of equal strength, remains free to reverse the process and restore the original polar field unless it is in some way eliminated. Babcock's proposal that it cancel with leader flux from the opposite hemisphere is attractive but is unsupported by observations.

Problems related to the reversals of the polar fields are further discussed in Chapter 10.

6.7 Forced oscillator models

LaBonte (see Wilson, 1987) and, independently, Bracewell (1988) argue strongly against relaxation (and any other) models in which the evolution of the large-scale surface fields determines the internal field configuration and the mechanism of the cycle. They have pointed out that, since the energy flux vector is directed outwards, and the energy densities of both the surface magnetic and velocity fields are several orders of magnitude less than those of their counterparts at, say, the base of the convection zone, it is unreasonable to expect that the former could seriously influence the evolution of the latter.

This argument has considerable force and provides a formidable objection to those versions of the Babcock–Leighton model in which the two-dimensional flux-transport equation is used to model the reversal of the polar surface fields. It is, indeed, hard to understand how purely surface interactions can give rise to the global sub-surface poloidal fields required by relaxation models.

Because of these difficulties, several writers, of whom the first was Walén

(1949), have explored the concept of the magnetic oscillator, in which the observed oscillating large-scale field is regarded as a temporary amplification of a primordial field by an oscillation in the differential rotation rate of the solar interior.

In their original form, such models were flawed because those simple enough to be solvable did not exhibit a discrete spectrum of oscillations. Adjacent oscillating shells were weakly, if at all, coupled, and hence the shells oscillated almost independently. By including interactions with the star's rotation, or identifying a preferred shell whose motion dominates the production of toroidal flux, a discrete spectrum could be achieved. In the latter case, the oscillator's period is determined by the period of the preferred shell. Layzer *et al.* (1979) identified the interface between the convection zone and the radiative interior as the location of this shell.

Bracewell (1988) has also proposed that the cycle is driven, or at least monitored, by a forced torsional oscillator located below the convection zone, which propagates 'magnetic waves' outwards through the convection zone to the surface. He envisages the presence of a clock, or flywheel, executing torsional vibrations, the period being set by a balance between the rotational kinetic energy of the flywheel and the potential energy of the stretched field lines. This interaction releases waves of azimuthal fields in dual conical beams at 11-year intervals at the base of the convection zone and, with considerable ingenuity, Bracewell suggests that refraction of the beam through the convection zone would cause it to impinge on the surface and generate the activity phenomena, first at higher latitudes and subsequently towards the equator.

Piddington (1972) has also been a vigorous critic of dynamo models, stressing the difficulties they encountered in eliminating the toroidal fields of the cycle by turbulent diffusion. He proposed an alternative oscillator model in which the angular velocity increases radially outwards, being (approximately) constant on concentric cylinders, and postulated that deep, rather than shallow, primordial poloidal field lines must, in the mean, lie quiescently along these cylindrical surfaces. Piddington argued that, if the poloidal field lines execute radial oscillations about this mean, with maximum amplitude in the equatorial plane and a period of 22 years, the field lines would be stretched into toroidal bands, within which loops or kinks would form and float upwards, giving rise to bipolar active regions, as in the Babcock model. Further, when the field lines arched upwards, the toroidal field would be of one sense, depending on the assumed sense of the poloidal field; when they were depressed downwards, the toroidal field would be in the reverse sense, thus satisfying Hale's polarity law for adjacent cycles.

Piddington's model assumes that, during the declining phase of each sunspot cycle, the toroidal field for that cycle is expelled from the photosphere and drifts off into the solar wind. It is thus unnecessary to invoke annihilation of the old field by turbulent diffusion. Further, the model could account for the overlap of the wings of the butterfly diagram by postulating that the expulsion process continued at low latitudes as the winding process of the next cycle was beginning, although why the results of this winding should first appear at high latitudes is unclear. Piddington's model was, however, unable to provide any account of the reversal of the observed polar poloidal fields.

Rosner and Weiss (1992) consider the torsional oscillations of a star using, for simplicity (but not realism), a cylindrical geometry. They assume a primordial field of 1 G, a shell of thickness $10^{-4}R_\odot$, a density of 10 gm cm^{-3}, and a shell radius of 5×10^{10} cm, and obtain an oscillator period of ~ 7.5 years, claiming that the result is 'plausible'. Indeed it is, but, in view of the uncertainties in the model and in the estimates, it is not conclusive.

The outstanding difficulty for all oscillator models is the matter of energetics. No one, so far, has been able to propose a mechanism whereby the nuclear energy of the core is harnessed to maintain the oscillator against the inevitable frictional damping and, given our present picture of the radiative zone and the core, it is difficult to imagine any physical process (or combination of processes) which could drive such a system. The question of the location of the oscillating shell can, at this stage, be answered only by guesswork or fiat; there is no evidence in the surface motions to support the existence of interior oscillations of this kind. There may be some evidence in recent helioseismology data (see Chapter 12) for oscillations within the radiative zone (Goode and Dziembowski 1991), but the accuracy of the data which relate to these regions is questionable.

It would seem that forced oscillator models create more difficulties than they resolve and, again, we are forced to look elsewhere for an explanation of the mechanics of the cycle.

6.8 Dynamo wave models

The driving mechanism of the Babcock model is the observed (surface) differential rotation, characterized by positive values of the differential coefficient $d\omega/d\theta$, where ω is the mean angular velocity at the co-latitude θ. In 1955, however, Parker (1955, 1957) assumed a positive radial gradient in the angular velocity $d\omega/dr$ and inferred from hydrodynamic theory the presence of a dynamo wave within the convection zone, consisting of azimuthal bands

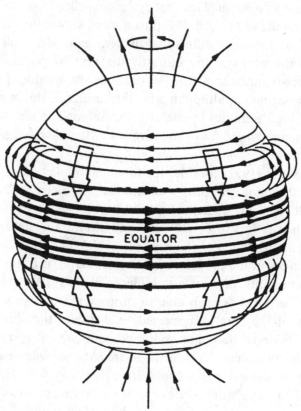

Fig. 6.3 The toroidal and poloidal fields associated with the dynamo wave model, as proposed by Parker (1957).

of toroidal magnetic fields alternating with a poloidal field which is 90° out of phase with the toroidal field. This system is illustrated schematically in Figure 6.3.

Observations of secondary magnetic phenomena (see §8.3) suggest that there should be two (or at most three) toroidal bands in each hemisphere at any one time and that 22 years would be required for the migration of each band from pole to equator. From dynamical considerations, Parker argued that the initial amplification of the relatively weak wave starting at the pole is primarily of the poloidal components, but, by the time that the wave reaches mid-latitudes, the decrease of cyclonic motions and the increase of the non-uniform rotation shift the amplification to the toroidal field. Although Parker was, at that time, unaware of the reversal of the polar fields, the fact that the polarity of the highest-latitude poloidal fields would change with each 'wave' implies that this model is not inconsistent with the reversals.

The butterfly diagram requires that, if a dynamo wave is responsible for the surface magnetic phenomena, it must propagate towards the equator in each hemisphere, at least at lower latitudes. Although Parker's original consideration of the direction of propagation has now been subsumed into a more mathematical discussion (see Chapter 11), it may be helpful to examine it here, because it highlights the essential features and some of the difficulties of what is now known as the '$\alpha - \omega$' dynamo (see § 11.3).

Assuming that $d\omega/dr$ is positive, Parker argued that a poloidal field in the meridional plane would be deformed into elements of a toroidal field in such a way that a positive poloidal element in the northern hemisphere would give rise to a positive (anti-clockwise, as seen from the pole) toroidal element, and a negative poloidal element to a negative toroidal element (see Figure 6.3). This is the 'ω-effect'. Under the action of cyclonic convection, loops or kinks then form in the toroid and float upwards, twisted by the Coriolis force as they rise, thus enhancing the local poloidal field (the 'α-effect'). If one assumes that the sense of rotation of the rising element is the same as in the core of the earth (i.e. anti-clockwise or positive in the northern hemisphere), the rising loop, characterized as an Ω-loop, should enhance each element of the poloidal loop on the equatorward side and cancel with it on the poleward side, thereby shifting the loop equatorwards.

On the other hand, Krause and Rädler (1980) have argued that, if the vertical dimension of a convective cell is more than one scale-height, the expansion of the rising cell should provide the dominant effect, and the sense of cyclonic rotation should be reversed. The poloidal loops should thus be enhanced on the poleward side and the wave should tend polewards. In this case, an equatorwards propagating dynamo could operate in the Sun only if $d\omega/dr$ were negative. Similar results are derived (below) from more formal theory, and the direction of propagation of a dynamo wave remains a difficulty for these models, constituting what Parker (1987) has called the 'dynamo dilemma'.

6.9 Conclusion

It would seem that, even in terms of the cyclic phenomena cited so far, none of these models ('the usual suspects') is satisfactory. The lack of any plausible driving mechanism is a major defect of the forced oscillator models, while both the Babcock model and the dynamo wave model face unresolved difficulties. We now discuss current data regarding stellar activity and activity cycles, followed by more recent solar observational data, which will lead to a more searching examination of the dynamo and other mechanisms.

References

Babcock, H. W.: 1961, *Astrophys. J.* **133**, 572.

Bracewell, R. N.: 1988, *Solar Phys.,* **117**, 261.

DeVore, C. R., Sheeley, N. R., and Boris, J. P.: 1984, *Solar Phys.,* **92**, 1–14.

DeVore, C. R., and Sheeley, N. R.: 1987, *Solar Phys.,* **108**, 47–59.

Giovanelli. R. G.:, 1985, *Australian J. Phys.,* **38**, 1045.

Goode, P. R., and Dziembowski, W. A.: 1991, *Nature,* **349**, 223.

Krause, F., and Rädler, K. H.: 1980, *Mean-field Magnetohydrodynamics and Dynamo Theory,* Pergamon, Oxford.

Layzer, D., Rosner, R., and Doyle, H. T.: 1979, *Astrophys. J.,* **229**, 1126.

Leighton, R. B.: 1964, *Astrophys. J.,* **140**, 1547.

Parker, E. N.: 1955, *Astrophys. J.,* **122**, 293.

Parker, E. N.: 1957, *Proc. Nat. Acad. Sci. USA,* **43**, 8.

Parker, E. N.: 1984, *Astrophys. J.,* **281**, 839.

Parker, E. N.: 1987, *Solar Phys.,* **110**, 11.

Piddington, J. H.: 1972, *Solar Phys.,* **22**, 3.

Rosner, R., and Weiss, N. O.: 1992, in *Proceedings of the National Solar Observatory/Sacramento Peak 12th Summer Workshop,* ed. K. L. Harvey, San Francisco, California, 511.

Sheeley, N. R., Nash, A. G., and Wang, Y.-M.: 1987, *Astrophys. J.,* **319**, 481.

Spruit, H. C., Title, A. M., and van Ballegooijen, A. A.: 1987, *Solar Phys.,* **110**, 115.

Walen, C.: 1949, *Ark. f. Mat. Astr. och Fys.,* **33A**, No. 18.

Wang. Y.-M., Nash, A. G., and Sheeley, N. R.: 1989, *Science,* **245**, 861.

Wilson, P. R.: 1987 *Solar Phys.,* **110**, 1.

Wilson, P. R., McIntosh, P. S., and Snodgrass, H. B.: 1990 *Solar Phys.,* **127**, 1.

7

Stellar activity and activity cycles

Scintillate, scintillate, globule vivivic
Fain would I fathom thy nature specific.
Variation on a well-known theme

7.1 The solar–stellar connection

Hale's conviction that solar physics is an essential component of astrophysics was shared by some, but unfortunately not by all, astrophysicists. Nevertheless, a further important initiative in this spirit was taken at Mount Wilson in 1966, when, using the 100-inch Hooker telescope, Olin Wilson and his colleagues began a long-term study of 91 cool dwarf stars (Wilson 1978). In 1978 this project (the 'HK Project') was transferred to the Mount Wilson 60-inch telescope, which was dedicated entirely to the continuation and extension of this work and has become identified with a particular methodology known as the *solar–stellar connection.*

Although the Sun permits the detailed two-dimensional study of its activity phenomena, it exhibits only a single set of stellar parameters, since its mass, size, composition, and state of evolution are necessarily fixed at this point of time. On the other hand, stars, as observed from earth, are essentially one-dimensional objects, but they offer a wide range of physical parameters which permit a more thorough testing of theories and conjectures regarding common phenomena than is possible for the Sun alone. The solar–stellar connection aims to bring these two lines of investigation together in order to further our understanding of the properties of the Sun and other late-type stars.

Even before the solar–stellar connection methodology became recognized as such, there were many examples of the successful application of solar results to stellar investigations. In the 1940s and 50s, for example, the chromosphere and corona were thought to be unique to the Sun, but, with the application of improved techniques from space research, equivalent regions have since been identified in the atmospheres of many stars.

On the other hand, studies of stellar evolution enable us to trace the past history and chart the probable future evolution of the Sun, yielding information unobtainable from even the most detailed contemporary knowledge of its present state of evolution. Studies of young, pre-main-sequence T Tauri stars provide insights into how the Sun, and possibly the solar system, evolved out of the solar nebula. Stars which have evolved off the main sequence into a red-giant phase indicate that the Sun, too, will eventually begin an expansion phase which will engulf the inner planets, some 5×10^9 years into the future.

The study of cyclic activity is another important area where a knowledge of stellar cycles can lead to both a greater understanding of the solar cycle and predictions on a more modest time-scale. Two general properties of stars will feature prominently in this study: the rotation rate (or period), and the presence and depth of a convection zone (see § 3.1). The main source of data on stellar rotation rates is high-resolution spectroscopy, which yields estimates of the projected rotation speed, $v \sin i$, from the analysis of spectral-line shapes. Rotation rates may also be inferred from short-period variations in luminosity, CaII emission, or any other indicator of activity arising from a non-uniform distribution of regions of activity on the stellar surface.

The presence of convection zones in stars was first suggested by studies of solar granulation, which implied the existence of a shallow convection zone in the Sun. The advent of sophisticated stellar modelling techniques indicated that, when the actual temperature gradient in any region exceeds the adiabatic gradient, convective heat transport should set in and a strictly radiative model must fail.

Any stellar model must, of course, involve some degree of uncertainty, but it is now generally accepted that the convection zone of the Sun extends some 200 000 km, or $\sim 30\%$ of the solar radius, below the surface. The existence and locations of stellar convection zones may also be inferred from stellar modelling techniques. Although the numerical accuracy of these techniques may not be high, the results are qualitatively important. In the very hot early-type stars of spectral types O and B, radiation is the dominant transport mechanism, except in the core itself. Shallow convection zones, beginning just below the surface, occur in models of stars of spectral type F, and these zones increase in depth (with respect to radius) through classes G, K, and M, which represent progressively cooler stars, until we reach mid- to late M-type stars, which are almost entirely convective.

Since the establishment of the Harvard classification system (O, A, B,... see Chapter 2) the increasing complexity of stellar spectra as perceived by modern high-resolution spectrographs has led to an explosion of sub- and sub-sub-classes and, to the uninitiated, the nomenclature associated with these classes is bewildering. In this chapter, many of these class names must, of necessity, be used without detailed explanation, but the general reader in search of further enlightenment is referred to any of the standard texts, e.g. Aller (1963).

Reference has been made in the preparation of this chapter to excellent review articles by Wilson, Vaughan and Mihalas (1981), Giampapa (1987, 1990), Radick, Lockwood and Baliunas (1990), Beasley and Cram (1990), Haisch, Strong and Rodonò (1991), and Radick (1992). New results presented at recent conferences and workshops are also reported, but it must be stressed that many of the data presented here have wide error bars and that the generalizations which have so far been attempted should be treated with some caution. The solar–stellar connection is, nevertheless, an exciting and potentially rewarding field of study. We begin with a review of the detection of solar-type phenomena in stars.

7.2 Stellar magnetic fields

Since the discovery of solar magnetic fields, it has been widely assumed that magnetic fields are also present in at least some classes of late-type stars. Only in recent years, however, have advances in spectral techniques permitted direct observation of stellar magnetic fields. Most positive detections are based on variations of a method, first described by Robinson, Worden, and Harvey (1980), in which the Zeeman signature in the intensity profile of a magnetically sensitive line is identified by comparison with that of a magnetically insensitive line.

Applications of this method have confirmed the presence of magnetic fields in a number of stars. Field strengths of $\sim$ 1–2 kG with area coverages of 20–80% are generally reported. While the field strengths are not surprising, this area coverage is large compared with that of the Sun ($\leq$ 1%). The higher degree of activity exhibited by these stars indicates that the difference between highly active stars and relatively quiet ones, such as the Sun, appears to involve the fractional area covered by magnetic fields on the stellar surface (the *filling factor*).

Coverages lower than 20% cannot, however, be detected by these methods. Consequently, if the Sun were observed as a star at a distance typical of even the nearer stars, its magnetic fields would have escaped notice.

7.3 Starspots

The possibility that luminosity fluctuations in stars of order 20% over periods of days, or a few weeks, might indicate that the surface of such stars was not uniformly bright was first suggested more than a century ago by Pickering (1880), who assumed that the star's rotation carried these variable or 'spotted' regions on and off the visible hemisphere, giving rise to a brightness or colour modulation of its continuous spectrum. Pickering tended to apply this hypothesis somewhat indiscriminately, and it was not until the middle of this century that Evans (1959, 1971) developed a detailed geometric spot model for the star CC Eri. Since 1970, many people (see Vogt 1980) have carried out photometric studies of a variety of stars, including the RS CVn subgiants, such as II Peg, which occur in binary pairs, and the BY Draconis dwarf stars (with luminosities $< L_\odot/2$), such as AU Mic, and have developed geometrical models in order to explain the light and colour variations. They find that the *starspot model*, usually invoking circular spots, is the only hypothesis which can successfully account for the wide diversity of photometric variations in these stars.

The term 'starspot' is perhaps unfortunate, since it may imply that the mechanisms which give rise to these variations are similar to those which produce sunspots. It should therefore be emphasized that, because of their extreme activity and the role played by tidal coupling, the RS CVn binaries are probably not particularly useful as guides to solar-type activity. Although the young BY Dra stars often appear to be single stars, they are, in fact, rapidly rotating low-mass dwarf stars, characterized by unusually intense chromospheric emission as well as by brightness variations of several per cent. The 'spots' produced on these stars may also be very different from sunspots: when a large sunspot group crosses the visible hemisphere of the Sun, the observed variation of the solar integrated flux is less than 1%, compared with up to 30% for the RS CVn and the BY Dra stars. The sensitivity of the ground-based stellar photometry used in these surveys is not sufficient to detect solar-type spots on stars, and the properties of 'starspots' detected by photometric methods differ in a number of ways from those of sunspots.

In the early stages of these studies, there was some doubt as to whether starspots were hotter or cooler than their surroundings, and, more recently, Petersen, Hawley, and Fisher (1992) have reiterated these concerns. However, detailed studies of the shape and size of both the light and colour variations have yielded typical 'effective' spot temperatures near 3400 ± 200K, generally $1000 - 1200$K cooler than the unspotted regions. If starspots have an umbra–penumbra structure similar to sunspots, where typically 97% of the flux comes from the penumbra, this result would imply umbral temperatures of $\sim$ 2600K. Starspots are therefore both actually and relatively cooler than sunspots, in which umbral and penumbral temperatures of 4000K and 5300K must be compared with a photospheric temperature of 5800K.

Once the starspot temperature is established, detailed reconstructions of spot sizes, shapes, and positions can be attempted, although it is not possible to claim uniqueness for such spot models. Spots reconstructed in this way tend to be extensive (up to 40% disk coverage), the light variation of II Peg being of the order of half a magnitude. The presence of polar spots is frequently inferred, although Giampapa (private communication) has challenged this interpretation of the data. Starspot lifetimes may exceed several years. For spotted stars, accurate determinations of rotation period (usually short, a few days) are possible, and, in a few cases, estimates of differential rotation have been attempted for younger stars (see Vogt 1983). The degree of spottedness is variable with time, with hints that spot activity on some late-type dwarfs is cyclic. Hartmann *et al.* (1979) have found evidence for slow changes in the mean luminosity on time scales of 30–60 years, indicative of cycles in the spot formation rate. The clearest evidence is on the star BD +26°730, which seems to show smooth and well-defined variations with a periodicity indicative of a 60-year spot cycle.

It has also been inferred that, unlike the Sun (see §7.12), an *increase* in spottedness in stars with large spots appears to correspond to a *decrease* in the mean energy output (Vogt 1975). It is therefore far from obvious that our knowledge of the structure of solar spots, such as it is, can be reasonably extrapolated to these 'monster' spots, or vice-versa.

7.4 Stellar flares

On 25 September 1948, a spectrogram of the dMe dwarf star L726-8, obtained by Joy and Humason (1949), revealed both line emission and continuum enhancement amounting to a change of about 1 magnitude. Thus began a long period of photometry of dKe/dMe UV Ceti-type stars in the U- and B-bands. Sudden brightenings on time scales ≤ 1 day have since been detected in the photometry and in the Hα and CaII spectra of many stars. By analogy with the spectral properties of solar flares, these brightenings have been interpreted as *stellar flares*.

These photometric studies concentrated on flare occurrence versus energy and the temporal evolution of the flare (Gershberg 1967, Moffatt 1974). Stellar radio flares also turned out to be unexpectedly bright (Lovell, Whipple, and Solomon 1963, 1964), but photometric flare studies languished until the mid 1970s, when interferometric techniques and the Very Large Array came into use. Not until the EUV/X-Ray window finally opened on the stellar side to reveal the high-energy stellar flare (Haisch *et al.* 1977) did solar and stellar studies of flares emerge from mutual isolation and begin their current trend of cross-fertilization as part of the solar–stellar connection.

If the Sun is typical, G dwarf stars should emit flares, but, because of the low optical contrast of a flare against the disk-integrated flux, it took the Exosat satellite to observe a large flare ($L_* = 10^{30}$ erg s^{-1}) on the active, rapidly rotating G0 V star π^1 U Ma (Landini *et al.* 1986). Other types of stars are likely to flare, but, unless the intensity of such flares is at least an order of magnitude greater than that of the typical solar flare, detection is problematic. Petersen (1989) has reviewed flaring activity across the entire H–R diagram, and Schaeffer (1989) has presented a compilation of 'flashes' ranging up to 7 magnitudes on several dozen stars. The dwarf K and M stars, the pre-main sequence T Tauri and FU Orionis stars, and the RS CVn stars are, however, more appropriate for this study.

Programs seeking to correlate flare activity with age, by comparing field flare stars with cluster flare stars (Tsvetkov and Tsvetkova 1990), have recently commenced. The most distant catalogued field dMe flare star is at 21 parsec, while the Hyades cluster is twice as distant; the Pleiades are at 127 parsec and the Orion associations at about 400 parsec. These low-mass stars evolve slowly: the approximate main-sequence lifetimes are 4×10^{10} years for a dK5 star, 7×10^{10} years for a dM0 star, and 27×10^{10} years for a dM5 star. Several hundred flaring objects have been catalogued in Orion and the Pleiades, studies by Mirzoyan *et al.* (1988) indicating that the field and cluster flare stars do represent the same class. Mirzoyan *et al.* also estimate that the galaxy contains 4.2×10^9 UV Ceti-type flare stars, which probably originated in now broken-up clusters and associations.

Pre-main-sequence stars show high levels of magnetic activity and strong flares (Fiegelson, Giampapa and Vrba 1990). FU Orionis stars may be in a phase between T Tauri and post-T Tauri stars, the entire sequence corresponding to the disappearance of circumstellar material. X-ray and radio flares in stars in the ρ Oph and other dark clouds show flares as energetic as those in the RS CVn systems. Some of this variability may be due to changes in obscuration and mass infall, but magnetic interactions between the disk and the star may also be occurring (Kuijpers 1989).

Radio mapping of the Pleiades shows no evidence of the giant radio flares reported in star-forming regions (Bastian, Dulk, and Slee 1988), which implies that, whatever the cause, the mechanism lasts less than a few $\times 10^7$ years.

An X-ray flare was observed on van Biesbrock 8, one of the faintest known dwarfs, a dM7e with $L_* \approx 5 \times 10^{-4} L_\odot$, and assumed to be fully convective (Tagliaferri, Doyle, and Giommi 1990), a result which is important in relation to the location of the dynamo.

This brief survey of stellar flares scarcely does justice to the extensive data available on this topic but is sufficient to the theme of this study. The interested reader is referred to the review 'Flares on the Sun and other Stars' by Haisch, Strong, and Rodonò (1991) for a more detailed survey.

7.5 Activity indicators

While some spectroscopic and photometric phenomena known to be associated with solar activity (e.g. X-rays, radio emission, photospheric variability) have been observed in stellar spectra and interpreted by analogy with solar phenomena (e.g. magnetic fields, starspots, and flares), detection is possible only where the degree of the inferred activity is substantially higher in the star than in the Sun. It would appear that the Sun is a very ordinary star and that any solar activity phenomenon may be present *a fortiore* in some other stars. It should therefore be remembered that there is a strong bias in studying activity in stars selected by these indicators and that inferences about solar conditions, based on the study of activity in such stars, might prove unreliable.

Fortunately, a few indicators of activity are sufficiently sensitive to allow us to explore stellar activity levels which are comparable with those found in the Sun. Some EUV lines satisfy this requirement: Hα and He λ 10830 provide important complementary data, but the emission cores of the H and K lines of CaII, and, more recently, of MgII, are the most widely used common indicators of stellar and solar activity. Strong correlations between the strengths of these lines in a given sample of stars suggest the presence of a common underlying factor, such as magnetic field. The correlation between the CaII emission and solar magnetic fields has, indeed, been well established (e.g. Skumanich, Smythe, and Frazier 1975).

The development of the 'H–K photometer' by A. Vaughan, as part of the HK Project, permitted the rapid sampling of the chromospheric emission of a star (usually less than 10 minutes per star) as measured by the H–K flux index. This index is given by the ratio of the number of photons passing through slits aligned with the H and K emission lines to the corresponding fluxes through wider slits away from the line cores. This instrument can obtain H–K flux measurements with a reproducible accuracy of 1–2%, despite the effects of the Earth's atmosphere and the small motions of the star's image.

In order to determine the star-to-star variations in the mean level of CaII, a one-off survey of 396 solar-type stars was included in the HK project, and, in Figure 7.1 the H–K flux index is plotted against the 'B–V' or colour index of the star. The latter represents the difference between the magnitude of the star when measured in the blue region of the spectrum and that when measured in the visual (yellow) region. Thus the hotter, blue stars lie on the left of the plot and the cooler, red stars on the right.

The plot shows a clear overall trend: the H–K index increases as the stellar temperature decreases. This does not necessarily indicate an absolute increase in the strength of the H and K emission, since most of it is due to the decrease in the continuum emission of the cooler stars. Relatively, however, the chromospheric emission increases significantly in cooler stars.

There is a considerable spread of points at any value of the B–V index. Any vertical cut on the diagram identifies stars of similar mass but different ages, with the youngest stars at the top of the cut and the oldest at the bottom. Although one might have expected a relatively smooth distribution of H–K flux indices of the stars along such a cut, there is a distinct gap, particularly towards the left-hand side, which is known as the Vaughan–Preston gap. Indeed, the whole diagram appears to bifurcate into two roughly parallel branches. Although it is conceivable that this is due to a hiatus in the birth of new stars at some particular epoch, it seems more likely that the mechanism which gives rise to chromospheric emission shifts abruptly from one mode to another, with a significantly reduced output as the age of the star increases.

7.6 Stellar activity, age, and rotation

In 1972, Skumanich (1972) stated his '$t^{-\frac{1}{2}}$' law for the time decay of stellar rotation and stellar chromospheric activity; i.e. the rotational velocity and the strength of the CaII emission of a late-type star vary inversely with the square root of the star's age. Durney (1972) provided a theoretical framework and derivation for these relations, and, for many years, these results were widely accepted. Since then, however, two fundamental discoveries have raised a serious challenge.

Vogel and Kuhi (1981) have shown that (i) low-mass, pre-main-sequence stars on their Hayashi tracks to the main sequence are, without exception, very slow rotators, and (ii) the rapid rotation speeds expected on the basis of extrapolations of the square-root relations are characteristic of only the

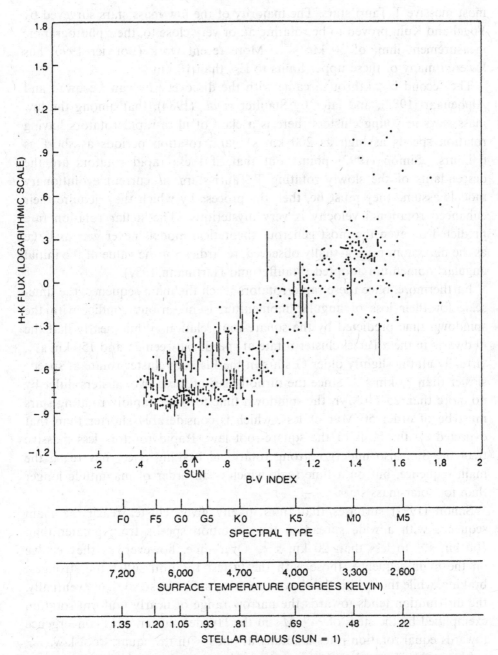

Fig. 7.1 The H–K flux variability for 91 stars in the Mount Wilson stellar activity survey (vertical bars) is compared with a 'snapshot' survey of nearly 400 stars (dots) representing a sample of all the sunlike main-sequence stars within 25 parsecs of the solar system in the northern half of the sky (from Wilson, Vaughan, and Mihalas 1981).

most massive T Tauri stars. The majority of the low mass stars surveyed by Vogel and Kuhi proved to be rotating at, or very close to, their photographic measurement limit of 30 km s^{-1}. More recent work (Bouvier 1990) has lowered many of these upper limits to less than 10 km s^{-1}.

The second breakthrough came with the discovery by van Leeuwen and Alphenaar (1982), and later by Stauffer *et al.* (1984), that among the low mass stars in young clusters there is a class of ultra-rapid rotators having rotation speeds as high as 200 km s^{-1} and rotation periods as short as 6 hours. Simon (1992) points out that, if these rapid rotators are the descendants of the slowly rotating T Tauri stars, as current evolutionary models assume they must be, then the process by which they acquire their enhanced rotational velocity is very mysterious. The stellar rotation rate predicted on even the most generous theoretical models never gets as large as the maximum rates actually observed, regardless of the value of the initial angular momentum adopted (Stauffer and Hartmann 1987).

Furthermore, once these rapid rotators reach the main sequence, the time-scale for their loss of angular momentum is in serious conflict with the spindown time predicted by the square-root relation. While nearly half the G dwarfs in the α Persei cluster rotate at speeds between 25 and 150 km s^{-1}, virtually all the slightly older G stars in the Pleiades cluster rotate at speeds slower than 25 km s^{-1}. Since the turnoff ages of these two clusters differ by no more than 25–50 Myr, the spindown time for such rapidly rotating stars must be of order 50 Myr or less, which is considerably shorter than that expected on the basis of the square-root law. Rapid rotators less massive than the Sun also undergo strong rotational braking once they reach the main sequence, but on a time scale which is an order of magnitude longer than for solar-mass stars.

Simon (1992) surmises that stars in any given cluster reach the main sequence with a wide spread in their rotation speeds from greater than 100 km s^{-1} to less than 20 km s^{-1}. Over time, however, as they evolve on the main sequence, those with the fastest rotation suffer the strongest braking, while those with the slowest rotation suffer least, so that, eventually, the distribution tends towards the narrow range of nearly uniform rotation exemplified by the stars of $\sim 1\,M_\odot$ in the Hyades cluster. This convergence towards equal rotation rates is totally at odds with the square-root law.

Simon also shows that stars in young clusters approach the main sequence with a wide range of chromospheric activity, as well as rotation rates. By the age of the Hyades, however, the most active stars have lost the greater part of their activity, while the least active have lost far less of their initial activity. Simon concludes that neither the spindown of individual stars as a function of age nor their activity–age relationship can be described by a single empirical formula such as the square-root law: these relations must be implicit rather than explicit. The more fundamental relationship, he argues, is between activity and rotation. According to Skumanich (1972) and Durney (1972) the strength of the CaII emission of a star is directly proportional to its surface (equatorial) velocity, and similar relationships have been proposed more recently for UV and X-ray emissions.

7.7 Activity and convection

Gray (1992a) has studied the presence of activity in giant stars of luminosity class III as they evolve across the H–R diagram on nearly horizontal tracks. Most leave the upper main sequence with rapid rotation, ≈ 140 km s^{-1}, and with a radiative envelope, but as they cross the 'granulation boundary' at $\sim$ F5 III, a convective envelope begins to develop and grows monotonically deeper for the rest of their trek across the diagram.

It is precisely when they cross the granulation boundary that dramatic atmospheric events become evident. The C II and C IV emission and related indicators show that the temperature inversion grows; the X-ray emission that joins these activities indicates the presence of coronal loops which were absent in slightly earlier evolutionary phases.

At this stage the convective envelopes are very thin, estimated to be only 3% of the stellar radius. A short evolutionary while later, the surface rotation rate of the giants decreases sharply, and the C II and C IV emission drops accordingly; several lines of evidence point towards strong magnetic braking. Among these are (i) the discontinuous nature of the decrease, which rules out simple moment-of-inertia changes, and (ii) the result that the distribution of rotation rates changes from a Maxwell–Boltzmann distribution in the stars before crossing the boundary to a single value after crossing it, an outcome which is consistent with the behaviour of the cluster stars discussed in §7.6.

At the other end of the scale, activity associated with spots and flares has long been observed in the cooler, less massive M dwarf stars, of which a fraction, the dMe flare stars, exhibit strong hydrogen line emission in addition to the H and K emission features. This evidence of hydrogen emission immediately indicates the presence of chromospheric activity, for the photospheres of these stars are too cool for such lines to form in their normal spectra.

Although small and cool, these stars are striking in that activity occurs on scales, both relatively and absolutely, greater than that of the Sun. Spots and other active features may cover 30% or more of an M dwarf's surface, and the giant flares occurring on dMe stars are 10 to 100 times more energetic than the largest solar flares and increase the star's luminosity by up to five magnitudes.

It is tempting to associate such intense activity with the extensive convection zones attributed to these stars. Although they are rather insignificant in their continuum emission, all M dwarf stars observed thus far show solar-type chromospheric emission, with sufficient spectral resolution to enable the detection of narrow emission cores at the bottom of the CaII H and K lines. It would therefore seem that the presence of an extensive convection zone in these stars is an essential ingredient. Although it has been suggested (e.g. Foukal 1990) that activity declines in stars later than M5, where the stellar interior becomes fully convective, this is not the case. Young, fully convective stars exhibit strong activity which decreases in older stars concurrently with a decrease in their rotational velocity (Giampapa, private communication).

7.8 The Rossby number

Since stellar activity is clearly influenced by both rotation rate and convection, theoreticians have looked for some particular combination of the two which might provide a more precise relationship. For solar-type stars, Noyes *et al.* (1984) found that the ratio of the rotation period to the turnover time of the largest convective eddies, known as the *Rossby number* N_R, provides that connection. Although there has been some criticism of this approach, there can be no denying that it works very well with the CaII emission and with every other chromospheric, transition region, and coronal diagnostic measured by the IUE and *Einstein* satellites (Simon, Herbig, and Boesgaard 1985). Figure 7.2, taken from Simon (1990), shows an example of such a rotation-activity curve for the chromospheric MgII lines. The rotation rate in stars with low Rossby numbers dominates the convective turnover rate, and it can be seen that a low Rossby number correlates remarkably well with strong mean MgII 1940 emission and strong surface activity. The closeness of this relation over an order-of-magnitude range of N_R for a wide range of spectral type, age, and activity levels in late-type dwarfs implies that the mechanism which determines the level of chromospheric emission, and therefore of magnetic activity, is similar throughout. A low value of N_R indicates a greater influence of the Coriolis forces arising from the effect of rotation rate on convection. As Chapter 11 demonstrates, this result implies an increasing importance of the α effect in the dynamo generation of the magnetic field.

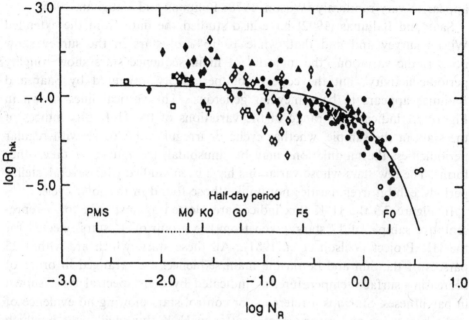

Fig. 7.2 The normalized MgII flux is plotted against the Rossby number: the squares denote pre-main-sequence stars, the diamonds represent cluster stars, the circles are late-type dwarfs, and the triangles are active chromosphere stars, i.e. spotted variables and RS CVn stars. The insert horizontal scale shows the Rossby number for stars with half-day periods (from Simon 1990).

7.9 Activity variability in individual stars

In addition to large star-to-star variations in the mean level of CaII emission, individual stars exhibit changes in the level of this emission over several time scales. The changes on the shortest time-scales (minutes–hours) are apparently related to stellar flares (see §7.4). The fairly regular modulations seen on intermediate time-scales (days–weeks) probably reflect the combined effect of rotation and longitudinal inhomogeneities. The rotation periods of a large number of stars have, in fact, been determined in this way, although there are several possible sources of error or misinterpretation (see Beasley and Cram 1990).

On even longer time-scales (years), the HK and related surveys have yielded a growing body of data concerning variations in the CaII emission. Of the 394 stars represented in Figure 7.1, 91 are part of the HK Project, with H–K flux emission monitored over periods of 10 years or more. Some of these stars exhibit fairly regular periodicities (5–20 years) and their activity variations appear to be analogous with those associated with the solar cycle.

Others vary irregularly, but with some discernible pattern, while yet others appear to vary randomly (Wilson, 1978; Baliunas and Vaughan, 1985).

Saar and Baliunas (1992) have also studied the data from the extended Wilson survey and find that, while 10–15% of stars in the survey show no periodic variation, 'the majority of main sequence stars show roughly periodic activity', but the concept of 'periodicity' favoured by Saar and Baliunas appears to be somewhat generous. The vertical lines shown in Figure 7.1 indicate the range of the variations of the H–K flux indices of the stars in the sample, whether cyclic or irregular. Of those with regular periodicities, the modulations may be sinusoidal, triangular, or take other forms. The few stars whose variations have been studied over several 'stellar periods' exhibit irregularities not unlike those found in the solar cycle.

In Figure 7.3 the H–K flux indices are plotted against time for a representative sample of 12 stars, taken from the group of 91 stars selected for the HK Project (Wilson *et al.* 1981). All these stars, which are within 25 parsecs of the Sun and lie on the main sequence, are arranged in order of decreasing surface temperature, as indicated by their spectral type, shown in parentheses. Star *a* is a reference or control star, showing no evidence of cyclic behaviour and a very small scatter in H–K flux, while star *b*, which is slightly hotter than *a*, shows a slow decline followed by a sharp increase. Star *c* is of the same spectral type as the Sun and exhibits a cycle of about 10 years. Although stars *d*, *e*, and *f* are all of type G8, they show marked differences in their behaviour, star *e* being an old star with low metallic content. Star *g* may be cyclic but has not yet been observed through a complete cycle, while star *h* exhibits rapid flux variations but no obvious cycle. Stars *i*, *j*, and *k* exhibit short, well-defined cycles, while star *ℓ* shows large variations but no obvious periodicity.

7.10 Characteristics of cyclically varying stars

The HK Project has shown that quasi-regular cyclic activity is by no means unique to the Sun, but it has also demonstrated that many active solar-type stars do not behave cyclically. If some stars exhibit solar-type cyclic behaviour, albeit with a range of periodicities, while other stars vary irregularly and yet others without any apparent pattern, then, by studying the different parameters of these stars (e.g. mass, luminosity, rotation periods, convective structure, age), it should be possible to identify those combinations of parameters most conducive to cyclic activity, and thus to understand the mechanism responsible.

Any attempt to associate stellar parameters with the occurrence of cyclic

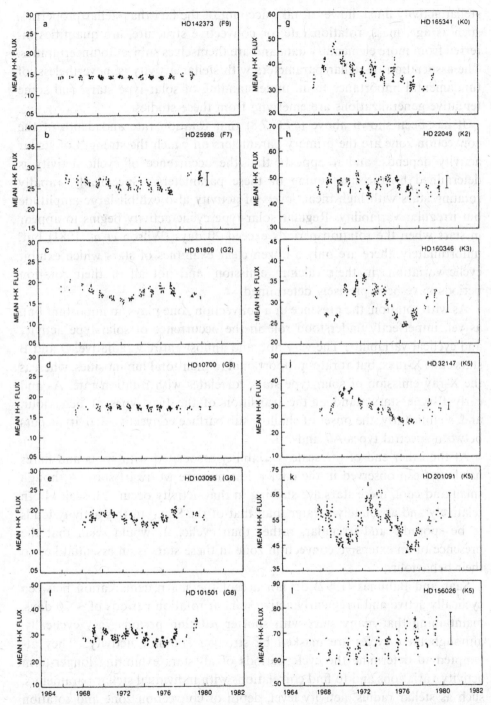

Fig. 7.3 The chromospheric mean H–K fluxes are shown for a representative sample of 12 stars selected from the group of 91 stars in the Mount Wilson stellar activity survey since 1986. All stars lie on the main-sequence and are within 25 parsecs of the solar system.

stellar activity must, however, take account of the fact that stellar properties, such as age, mass, rotation rate, or convective structure, are quantities inferred from more elementary data and are themselves subject to uncertainties. The association of stellar parameters with stellar activity is, nevertheless, of fundamental importance to an understanding of solar-type stars, and some tentative generalizations are emerging from these studies.

It has been shown above (§ 7.6–7.8) that rotation rate and depth of the convection zone are the primary parameters on which the strength of stellar activity depends, and it appears that the occurrence of cyclic activity is determined by the fine tuning of these parameters. Specifically, rapidly rotating stars with high mean levels of activity also exhibit large-amplitude but irregular variability. Regular solar-type cyclic activity begins to appear in stars when the rotation periods exceed 20 days (Wilson *et al.* 1981) but, unfortunately, there are only a dozen clear examples of stars which exhibit cyclic variations in their calcium emission, and not all of their rotation periods have been precisely determined.

As with rotation, the presence of a convection zone plays an important but, as yet, imperfectly understood role in the occurrence of solar-type activity and cyclical variability. The very hot, and almost entirely radiative, O and B stars emit X-rays, but at rates proportional to their total luminosities, whereas the X-ray emission of solar-type stars correlates with rotation rate. A- and early F-type stars stand at the transition of the two types of behaviour, and, significantly, the onset of shallow sub-surface convective activity occurs between spectral types A7 and F0.

At the other end of the scale, variability associated with spots and flares has long been observed in the cooler, less massive M dwarf stars. Although small and cool, these stars are striking in that activity occurs on scales both relatively and absolutely greater than that of the Sun, but this activity tends to be sporadic and irregular, rather than cyclic. It would seem that the presence of an extensive convection zone in these stars is an essential key to their behaviour.

Saar and Baliunas (1992) do not accept the sharp demarcation between cyclically active and irregularly active stars at rotation periods of ∼ 20 days, maintaining that many stars with shorter rotation periods vary cyclically, although their cycles are masked by stronger sporadic activity. They attempted to determine the cycle periods of all stars exhibiting longer-term activity variations and to find correlations with individual stellar parameters, such as stellar radius, activity level, depth of convection zone and rotation rate, but found no significant correlations in their sample. However, while their plot of cyclic period against rotation rate (their Figure 6) shows an apparently random spread of points, it may be significant that, if the stars with rotation periods less than 25 days are excluded, there appears to be a clear trend of increasing cycle period with rotation period.

At this stage, the sample size is small and the data string extends only over at most 25 years, so that the cycle periods of stars with rotation periods less than 20 days, if real, are likely to be poorly determined. It is well-known that the occurrence of activity on the Sun is irregular on time-scales of less than a year, and it may be that the 20-day rotation period marks the point beyond which the amplitudes of irregular activity variations sink below some critical level, so that the fundamental periodicity is no longer obscured.

Saar and Baliunas (1992) claim a correlation between the normalized cycle frequency, $\Omega^*_{cyc} = R^2 \Omega_{cyc}/\eta_M$, (here R is the stellar radius, Ω_{cyc} is the cycle frequency, and η_M the magnetic diffusivity) and a number related to the $\alpha - \omega$ dynamo (see §11.3), which they define by $N_{\alpha\omega} = d^{0.5} R^2 \Omega_{rot}/\eta_M$ (d being the depth of the stellar convection zone and Ω_{rot} the rotational frequency). For old, relatively inactive stars, such as the Sun and other G and K dwarfs, they find the relation,

$$\Omega^*_{cyc} \approx k N_{\alpha\omega}^{1.6}$$

(here k is a constant of proportionality), while younger, more active stars, such as HD154417, follow the relation

$$\Omega^*_{cyc} \approx k' N_{\alpha\omega}^{1.3}.$$

Noting that dynamo theory predicts a linear relation between Ω^*_{cyc} and $N_{\alpha\omega}$ (Tuominen *et al.* 1988), they caution that there is considerable uncertainty in their estimates of the parameters involved and that these correlations cannot yet be regarded as well-established.

At this stage it must be accepted that the onset of cyclically varying activity is not clearly defined, and that the relations between cyclic periods, rotation rates, and other stellar parameters for cyclically active stars are not well determined nor understood.

It is, nevertheless, reasonable to conclude that the occurrence of cyclic activity depends on a balance between convection and rotation. An adequate (i.e. neither too deep nor too shallow) convection zone is necessary, as is a rate of rotation that is not too rapid. When the convection zone is too shallow, activity is weaker and cyclic patterns are not detected; when the convection zone is too extensive or the rotation rate too rapid, cyclic activity is overcome by stronger and more sporadic activity. The significance of these results for dynamo models is discussed in Chapters 11 and 15.

7.11 Grand minima in stars

Although it may not be surprising that stars with stellar parameters similar
to those of the Sun exhibit solar-like variability in their CaII emission, it
is puzzling that some stars in the same stellar category do not. Baliunas
and Jastrow (1990) have studied the level of magnetic activity in some 70
solar-type stars, as measured by the ratio S of the emission in the Ca H and
K lines, averaged over time, to that in the neighbouring continuum. Their
histogram exhibits two peaks, which indicates two different populations, one
containing some 60% of the sample, which peaks at $S \approx 0.17$, and the other
(40%), which peaks at $S \approx 0.145$. They note that, during the solar cycle,
$S_\odot$ ranges between 0.164 and 0.178 and that, for zero magnetic activity,
$S = 0.14$.

Baliunas and Jastrow's detailed study of the variations of the CaII emission
of thirteen of these stars found that nine, belonging to the first population,
exhibited solar-type variability, whereas for the remaining four, belonging
to the second, calcium emission was essentially constant and less than the
minimum of the stars with variable emission.

From these results, they advance the ingenious hypothesis that a subset of
the latter population of stars is not permanently quiescent. Since these stars
are in other respects similar to the Sun, Baliunas and Jastrow suggest that
at least some are passing temporarily through periods of reduced activity
similar to the Maunder Minimum of the Sun. The fraction of time that the
Sun appears to spend in such intermissions is, they note, of the same order
as the ratio of the second population of stars to the total but, in each case, it
must be remembered that the sample sizes on which these comparisons are
based are small.

Until one of the stars in the second population is observed to transfer to
the first (or vice-versa), this hypothesis must remain pure speculation. The
proposition is, however, difficult to disprove and, to follow our rule-of-thumb
that every phenomenon observed on the Sun will eventually be found on a
star, it remains a very attractive speculation. If true, we may greatly extend
the time base of direct observations of the evolution of solar-like cycles
beyond the few hundreds of years of solar data.

7.12 Photospheric variability in active stars

Photometric studies of the Sun (see §5.6) have shown that both short-term
(seasonal) and long-term (cyclic) variations in the solar output accompany
the chromospheric activity variations, although the amplitudes of the photo-

metric variations are small (< 0.1%). Interestingly, although the seasonal luminosity variations are out of phase with the activity variations (being dominated by the blocking effect of sunspots), the cyclic variations appear to be in phase with the level of activity.

The first successful attempts to detect photometric brightness variability in single cool dwarf stars were undertaken about ten years ago by R. Radick and his colleagues in a collaborative project involving both the Lowell (Arizona) and Mount Wilson Observatories. In some cases, bright field stars were selected from Olin Wilson's survey; in others, members of nearby stellar clusters were observed (Lockwood *et al.* 1984, Radick *et al.* 1987). The most sustained of these efforts involves the Hyades, a young cluster about one-fifth the age of the Sun, and has shown that persistent brightness variations at levels of a few per cent on time scales of days or weeks are common among cool dwarf Hyades stars. As in the case of the Sun, rotational modulation plus evolving stellar magnetic activity adequately explain the observed short-term photometric variations.

Parallel photometric and chromospheric CaII H and K emission observations of several Hyades stars, made at the Lowell and Mount Wilson Observatories respectively, show two characteristic patterns: (i) the rotational modulation signals from the two observatories are out of phase; and (ii) abrupt, secular changes in photometric brightness are frequently accompanied by opposite, but more gradual, changes in chromospheric CaII H and K emission. These patterns suggest that the stellar surface markings comprise dark continuum features both spatially and temporally associated with bright CaII emission, a configuration strongly reminiscent of solar active regions with their dark sunspots and bright emission plages.

The Lowell Hyades observations also provide the first data concerning long term (year-to-year) brightness changes among young, solar-type stars. Like the short-term fluctuations, the longer-term brightness variations amount to a few per cent and are accompanied by changes in the mean chromospheric activity. In contrast to the Sun, however, these young stars become brighter as their mean level of activity decreases and dimmer as activity increases.

These observations raise several questions. Does the relation between longer-term photometric variability and chromospheric activity observed among Hyades stars characterize other young active stars, or is it peculiar to the Hyades cluster? How do older, less active stars behave? Is the Sun unusual in this respect?

To address such questions and to explore the regime of stellar variability below amplitudes of 1%, a new photometric study was initiated at the Lowell Observatory in 1984. Twenty-nine stars in the new Lowell study were selected from Wilson's original chromospheric activity survey; four more were selected from the Mount Wilson HK Project, bringing to 33 the total number of stars for which parallel observations were made. A variety of stars was included in the sample: young, active stars similar to the Hyades stars; older, less active stars with observed cyclic chromospheric activity; and the HK Project standard stars which showed little chromospheric variability.

Sixteen of the 33 stars in the new sample exhibit statistically significant, short-term (seasonal) photometric brightness variations over two or more years at levels ranging from 0.2% to over 1%, with rotational modulation a likely contributing factor. The onset of persistent, short-term variability occurs rather suddenly at a mean activity level of about twice the solar value and is confined largely to the younger, more active stars of the sample. Short-term photometric variability was not frequently detected among older, less active stars during the survey, a probable consequence of the threshhold imposed by night-to-night instrumental errors rather than lack of intrinsic variability among these stars. The Sun itself would not register a seasonal variability signal if observed to the same precision. Intercomparison of the results from the two observatories, together with the Sun's known photometric seasonal variation, led Radick *et al.* to conclude that all ordinary cool dwarf stars may exhibit seasonal photometric variations, that the sense of the short-term brightness and chromospheric emission variations may be the same for all such stars, including the Sun, and that active regions are characteristic surface features of all of them.

Radick *et al.* also looked for longer-term (year-to-year) photometric variability and found that 21 of 29 stars observed over a four-year period showed significant variations ranging from less than 0.5% to over 4%.

Significant year-to-year variations were detected among older, less active stars. Unlike seasonal variations, which reflect non-uniformities in the spatial distribution of their surface markings, the year-to-year variations appear to be due to real evolutionary changes in the number and contrast of these features. Thus, even though they are probably linked by a common underlying agency, such as magnetic fields, one does not necessarily cause the other, and statistical studies show that they are not functionally related. There is, however, an excellent correlation between the amplitudes of the short-term and longer-term variability in the sample, which suggests that stars tend to preserve irregularities in the distribution of their surface features even as the number of these features changes.

Unlike the short-term brightness variations, the sense of the relation between the longer-term photospheric and chromospheric variations depends on the mean level of the star's activity. The younger, more active stars in the sample show the sense of correlation which characterizes the Hyades stars: photospheric brightness varies inversely with chromospheric activity changes. On the other hand, older, less active stars tend to display solar-type brightness variations which are in phase with chromospheric activity levels.

7.13 The parameters of variability

In a more recent attempt to relate stellar parameters to stellar variability, Radick (1992) has investigated the variability of photospheric and chromospheric emission of 55 F-, G-, and K-type stars which are not binaries. In order to study the effects of rotation rate, he compared stars of comparable masses, i.e. convective turnover times. HD81809, a G2V star, has a rotation period of 20 days and exhibits a cyclic variation in its CaII emission of 8.3 years, which is roughly in phase with a luminosity variation comparable in its amplitude to that of the Sun ($\sim 0.1\%$). By comparison, HD1835, also a G2V star, has a rotation period of only 8 days and an irregular, non-cyclic variability in its CaII emission, which anti-correlates with luminosity fluctuations of a few per cent.

By selecting stars of similar rotation periods (~ 8 days) but different masses and convection zone characteristics, Radick showed that F-type stars with thin convection zones exhibit less photometric variability (continuum brightness variations of amplitude $\sim 0.1\%$) by comparison with G- and K-type stars of comparable rotation periods. Although they may exhibit strong activity, F-type stars show, at most, weak short- and long-term variability, an indication that they are unable to produce the localized active regions which would give rise to a rotational modulation of their CaII emission. The same general pattern is found among stars of longer (>20 days) rotation periods.

Radick (1992) has also studied the relationship between the continuum and the chromospheric brightness variations within the late F- through mid K-type populations and again found that the younger, more active, and more rapidly rotating stars show a negative correlation (i.e. if they are brighter in the chromosphere, they are weaker in their photometric output), whereas for older, less active, and more slowly rotating stars, the correlation is positive. He has also found that stars in the latter category tend to exhibit regular activity cycles, while those in the former do not. Radick interprets the brightness anti-correlation of the former by suggesting that younger stars tend to direct their 'activity' into the production of large, dark spots with few faculae and explains the positive correlation of the latter by assuming, on the basis of solar studies (e.g. Foukal, 1990), that the balance shifts as the star ages and slows, so that excess facular emission outweighs the reduced emission from the spotted regions of these stars on long-term (cyclic) time scales.

7.14 The stellar ageing process

It has been shown that the modes of stellar activity depend critically on rotation rate and convection. Although Simon (1990) has shown that the simple $t^{-\frac{1}{2}}$ law does not apply in many cases (e.g. pre-main-sequence stars, see §7.6), there is little doubt that, as main-sequence stars age, their rotation rate slows down, and, through this connection, activity phenomena on ordinary cool dwarf stars may be qualitatively related to the age of the star. On young stars these features are more numerous, or larger, or have higher contrast than on older stars such as the Sun. The dark spots on young stars must also be more prominent relative to the bright continuum features, since the spots appear to dominate the photometric brightness variability on all time scales for them, rather than on just the short time scales, as for the Sun. The observed amplitudes of the photometric variations suggest that the fraction of the stellar surface covered by dark spots is at least ten times greater than for the Sun, but the surface coverage of bright features is enhanced by a smaller factor of ~ 3. It would seem that young stars prefer to arrange their larger amounts of surface flux into dark spots or groups rather than into bright faculae and magnetic network features.

As a solar-type star ages, its rotation rate declines, its level of activity decreases, and it tends to develop regular, or cyclic activity habits, in contrast to the more sporadic and much more exuberant activity of younger stars. (It is hard to resist the comment that the parallel with the human ageing pattern is remarkably close.) The evolution of the star's brightness variability is probably more complicated, because the balance between the competing effects of dark spots and bright features, which initially favours the spots on all time-scales, gradually shifts, as the star ages, to favour the bright features on the time-scale of stellar activity cycles.

An overall decline in variability, reflecting the decline in stellar activity, also occurs. At some point, however, the spot deficit and the bright feature excess must pass through a stage of temporary balance. A solar mass star would reach that point after about 10^9 years, which is somewhat more than the age of the Hyades cluster. The mean level of activity would still be higher than that of the present Sun and would show considerable short-term brightness variations, arising from rotational modulation and active region evolution. When the star reaches the present age of the Sun, the bright feature excess has become dominant on cyclic time scales, but, in the short-term, variability appears to be dominated by the spots.

7.15 Line asymmetries in stars

A recent development in solar–stellar connection studies has been the observations of changes in spectral line asymmetries in both the Sun (Livingston 1991) and cyclically varying stars (Gray 1992b). Sharp decreases in the magnitude of the asymmetries have been noted at certain phases of the stellar cycles, and these are interpreted as indicating changes in the surface convective properties of the stars. Presumably the changes are due to changes in the magnetic flux, which modifies the convective patterns by at least partly suppressing the convective flows. In the Sun the largest asymmetries are observed near sunspot minimum, when the magnetic flux density and luminosity are lowest, and similar variations are found in the K0 dwarf star σ Dra. In contrast, the more active K2 dwarf ϵ Eri exhibits the largest asymmetries when the luminosity is greatest.

Saar and Baliunas (1992) suggest that the asymmetries are anti-correlated with luminosity in the Sun and in σ Dra, for the same reasons that activity and luminosity variations are correlated in such stars (§ 7.14); i.e. because the magnetic flux is greatest (and convection accordingly suppressed) when the surface coverage of bright plage is at a maximum. The more active ϵ Eri has a larger spot-to-plage ratio, and asymmetries are therefore correlated with luminosity, because for younger stars the magnetic flux is greatest when the luminosity is least, due to the large spot-filling factor.

7.16 Conclusion

It must be stressed that the sample sizes of the studies on which these generalizations are based are not large and that they should be treated with caution. There is, however, much to learn about the nature of solar and stellar cycles by studying the interrelation of these stellar parameters with chromospheric and photospheric variability in the activity patterns of F-, G-, K- and M-type stars. In particular, the obvious dependence of magnetic activity patterns on both rotation speed and depth of convection zone strongly suggests the presence of some form of dynamo action, the reasons for which are discussed in later chapters.

References

Baliunas, S. L., and Vaughan, A. H.: 1985, *Ann. Rev. of Astron. and Astrophys.*, **23**, 379.

Baliunas, S. L., and Jastrow, R.: 1990, *Nature*, **348**, 520.

Bastian, T. S., Dulk, G. A., and Slee, O. B.: 1988, *Astron. J.*, **95**, 794.

Beasley, A. J., and Cram, L. E.: 1990, *Solar Phys.*, **125**, 191–208.

Bouvier, J.: 1990, *Astron. J.*, **99**, 946.

Durney, B.: 1972, NASA SP-308, 282

Evans, D. S.: 1959, *Mon. Not. R. Astr. Soc.*, **119**, 526.

Evans, D. S.: 1971, *Mon. Not. R. Astr. Soc.*, **154**, 329.

Fiegelson, E. D., Giampapa, M. S., and Vrba, F. J.: 1990, in *The Sun in Time*, eds. C. P. Sonnett, M. S. Giampapa, and M. S. Matthews, University of Arizona.

Foukal, P. V.: 1990, *Sci. Am.*, **262**, 34.

Gershberg, R. E.: 1967, *Astrofizika*, **3**, 127.

Giampapa, M. S.: 1987, *Sky and Telescope*, **74**, 142.

Giampapa, M. S.: 1990, *Nature*, **348**, 448.

Gray, D. F.: 1992a, in *Proceedings of the National Solar Observatory/Sacramento Peak 12th Summer Workshop*, ed. K. L. Harvey, San Francisco, California, 472.

Gray, D. F.: 1992b, in *Cool Stars, Stellar Systems and the Sun*, ed. M. Giampapa and J. Bookbinder, *Pub. A.S.P.*, Conf Series, **26**, 127.

Haisch, B. M., Linsky, J. L., Lampton, M., Paresce, F., Margon, B., and Stern, R.: 1977, *Astrophys. J.*, **213**, L119.

Haisch, B., Strong, K. T., and Rodonò, M.: 1991, *Ann. Rev. of Astron. and Astrophys.*, **29**, 275.

Hartmann, L., Londono, C., and Phillips, M. J.: 1979, *Astrophys. J.*, **229**, 183.

Hartmann, L., and Rosner, R.: 1979, *Astrophys. J.*, **230**, 802.

Joy, A. H., and Humason, M. L.: 1949, *Pub. Astron. Soc. Pacific*, 133.

Kuijpers, J.: 1989, in *Solar and Stellar Flares*, IAU Colloquium 104, eds. B. M. Haisch and M. Rodonò, Kluwer, Dordrecht, Holland, 163.

Landini, M., Monsignori-Fossi, B. C., Pallavicini, R., Piro, L.: 1986, *Astron. and Astrophys.*, **157**, 217.

Livingston, W.: 1991, in *The Sun and Cool Stars: Activity, Magnetism, Dynamos;* IAU Colloquium 130, eds. I. Tuominen, D. Moss, and G. Rüdinger, Springer Verlag, Berlin, 246.

Lockwood, G. W., Thompson, D. T., Radick, R. R., Osborn, W. H., Baggett, W. E., Duncan, D. K., and Hartmann, L. W.: 1984, *Publ. Astron. Soc. Pacific*, **96**, 714.

Lovell, B., Whipple, F. L., and Solomon, L. H.: 1963, *Nature*, **198**, 228.

Lovell, B., Whipple, F. L., and Solomon, L. H.: 1964, *Nature*, **202**, 377.

Mirzoyan, L. V., Ambaryan, V. V., Garibdzhanyan, A. T., and Mirzoyan, A. L.: 1988, *Astrofizica*, **29**, 531.

Moffatt, T. J.: 1974, *Astrophys. J. Suppl.*, **29**, 1.

Noyes, R. W., Hartmann, L. W., Baliunas, S. L., Duncan, D. K., and Vaughan, A. H.: 1984, *Astrophys. J.*, **279**, 763.

Petersen, B. R.: 1989, in *Solar and Stellar Flares*, IAU Colloquium 104, eds. B. M. Haisch and M. Rodonò, Kluwer, Dordrecht, Holland, 299.

Petersen, B. R., Hawley, S. L., and Fisher, G. H.: 1992, *Solar Phys.*, **142**, 197.

Pickering, E. C.: 1880, *Proc. Am. Acad. Arts Sci.*, **16**, 257.

Radick, R. R.: 1992, in *Proceedings of the National Solar Observatory/Sacramento Peak 12th Summer Workshop*, ed. K. L. Harvey, San Francisco, California, 450.

Radick, R. R., Thompson, D. T., Lockwood, G. W., Duncan, D. K., and Baggett,
     W. E.: 1987, *Astrophys. J.*, **321**, 459–72.
Radick, R. R., Lockwood, G. W., and Baliunas, S. L.: 1990, *Science*, **247**, 39.
Robinson, R. D., Worden, S. P, and Harvey, J. W.: 1980, *Astrophys. J.*, **236**, L155.
Saar, S. H., and Baliunas, S. L.: 1992, in *Proceedings of the National Solar
     Observatory/Sacramento Peak 12th Summer Workshop*, ed. K. L. Harvey, San
     Francisco, California, 150.
Schaeffer, B. E.: 1989, *Astrophys. J. Suppl.*, **337**, 927.
Simon, T.: 1990, *Astrophys. J.*, **359**, L51.
Simon, T.: 1992, in *Cool Stars, Stellar Systems and the Sun*, ed. M. Giampapa and
     J. Bookbinder, *Pub. Astron. Soc. Pacific*, Conf. Series, **26**, 3.
Simon, T., Herbig, G. H., and Boesgaard, A. M.: 1985, *Astrophys. J.*, **293**, 551.
Skumanich, A.: 1972, *Astrophys. J.*, **171**, 565.
Skumanich, A., Smythe, C., and Frazier, E. N.: 1975, *Astrophys. J.*, **200**, 747–64.
Stauffer, J., Hartmann, L. W., Soderblom, D. R., and Burnham, N.: 1984,
     Astrophys. J., **280**, 202.
Stauffer, J., and Hartmann, L.: 1987, *Astrophys. J.*, **318**, 337.
Tagliaferri, G., Doyle, J. G., and Giommi, P.: 1990, *Astron. and Astrophys.*, **231**, 131.
Tsvetkov, M. K., and Tsvetkova, K. P.: 1990, in *Proc. IAU Symp.*, 137.
Tuominen, I., Rüdiger, G., and Brandenburg, A.: 1988, *Activity in Cool Star
     Envelopes*, eds. O. Havnes *et al.*, Kluwer, Dordrecht, Holland, 13.
van Leeuwen, F., and Alphenaar, P.: 1982, *ESO Messenger*, **28**, 15.
Vogel, S. N., and Kuhi, L. V.: 1981, *Astrophys. J.*, **245**, 960.
Vogt, S. S.: 1975, *Astrophys. J.*, **199**, 418–26.
Vogt, S. S.: 1980, *Astrophys. J.*, **240**, 567–84.
Vogt, S. S.: 1983, *Proceedings of the 71st Colloquium of the International
     Astronomical Union*, Catania, Italy, 10–13 Aug 1982, 137–56.
Wilson, O. C.: 1978, *Astrophys. J.*, **226**, 379–96.
Wilson, O. C., Vaughan, A. H., and Mihalas, D.: 1981, *Sci. Am.*, **244**, 89.

8

The two-dimensional representation of the extended activity cycle

For you'll look sweet upon the seat of a bi-cycle built for two.

Music-hall song

8.1 Introduction

Chapter 5 described the one-dimensional characteristics of the solar activity cycle in terms of variations of the scalar function $N(t)$, where N is a number representing the number of sunspots (or sunspot groups or faculae or any other indicator) and t is the time. The butterfly diagram, shown again in Figure 8.1, provides a two-dimensional characterization of the cycle in terms of a function $N(\lambda, t)$, where λ is the latitude, a representation which yields additional information of obvious importance to the heuristic models discussed in Chapter 6. (Strictly speaking, the existence of active longitudes emphasizes that N is a function of three variables, $N(\lambda, \phi, t)$, where ϕ is the Carrington longitude.)

It has been known for ~ 130 years that the wings of the butterfly diagram overlap to some extent. The overlap is marginal between some cycles, e.g. 18 and 19, but in other cases, e.g. 19 and 20, it extends over ~ 2 years. Sunspot minimum, therefore, is a point on the one-dimensional plot determined partly by the decay of the old cycle and partly by the rise of the new.

In the one-dimensional approach, all activity lying between successive minima is associated with that particular cycle but, in the two-dimensional approach of the butterfly diagram, active regions of each cycle are distinguished by two properties: latitude and orientation of the magnetic axis (see §2.10). Spots of the new cycle should appear at higher latitudes $(20° - 40°)$ and exhibit a reversed magnetic orientation compared with those of the old cycle for a given hemisphere. The overlap of the wings indicates that

122

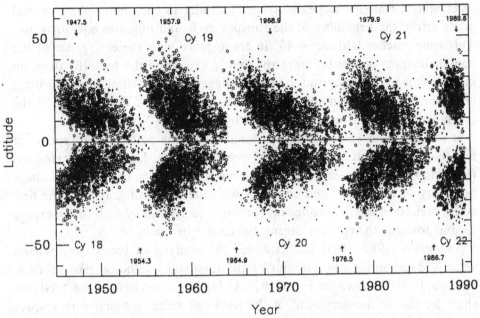

Fig. 8.1 The latitude distribution of sunspot regions against time (i.e. the butterfly diagram) from 1945 to 1991. The times of sunspot maxima and minima are indicated by vertical arrows at the top and the bottom of the diagram respectively. The overlaps of Cycle 20 over Cycle 19 and of Cycle 21 over Cycle 20 are obvious. (From Harvey 1992.)

sunspots from two adjacent cycles may be present simultaneously on the Sun. If this is the case, then the one-dimensional concept is inadequate.

Richardson (1948), Giovanelli (1964), and Dodson *et al.* (1982) used these criteria to identify the first and last spots of Cycles 14–20 from Greenwich and Mount Wilson data, and they found that the time between the appearance of the first and last spots of a given cycle ranged between 12.0 and 14.2 years, i.e. 1.8 to 3.3 years longer than the corresponding times between successive minima.

These and other data are examined in this chapter to determine whether this overlap should be expanded into the concept of an *extended activity cycle*, and the significance of this concept for models of the cycle is assessed.

8.2 Other indicators of activity

Coronal observers in the 1950s (e.g. Waldmeier 1957, Trellis 1957, Bretz and Billings 1959) noted that, in addition to the strong maxima in the emission of the 5303 Å 'green' emission line of Fe XIV at sunspot latitudes, a zone of enhanced emission appears at high latitudes in each hemisphere several years before the beginning of the sunspot cycle and migrates equatorwards. This zone reaches latitude $\sim 40°$ in coincidence with the emergence of the new cycle spots at this latitude, apparently extending the butterfly diagram back in time to higher latitudes. Similar suggestions regarding a high-latitude component have been made on the basis of a study of the variance of the green-line emission from 1944 to 1974 by Leroy and Noens (1983).

It was noted in § 3.4 that the *torsional oscillation* pattern first appears at high latitudes and tends progressively equatorwards. LaBonte and Howard (1982) found that, as the bands enter sunspot latitudes, the poleward shear boundary of the faster band coincides with the centre of gravity of the first active regions of the new cycle and continues in synchrony with the emerging regions towards the equator during the course of the cycle.

Snodgrass (1987, 1991) has extended the analysis of the Mount Wilson velocity data, and a contour plot of the torsional oscillation pattern from 1968 to 1990 is shown in Figure 8.2(a). He has also derived the torsional shear by taking the derivative of the torsional velocity signal with respect to latitude, and his contours of greater-than-average and less-than-average shear are plotted on the synoptic diagram in Figure 8.2(b). The greater-than-average and less-than-average shear contours are compared with the sunspot butterfly pattern in Figures 8.3(a) and 8.3(b) respectively.

Three branches of the greater-than-average shear pattern may be found in each hemisphere. The first appears at low latitudes between 1968 and 1979, Figure 8.3(a) demonstrating that it closely matches the butterfly diagram for Cycle 20. The second can first be identified at high latitudes between 1968 and 1972 in Figure 8.2(b), and Snodgrass notes that this polar spin-up in the net torsional pattern occurs at the time when the polar fields are reversing. In 1972 this component of the pattern begins a steady progression to lower latitudes, where it follows the butterfly pattern for Cycle 21 (see Figure 8.3(a)) to the equator, where it ends in 1988 (see Figure 8.2(b)). The third branch can be seen at high latitudes in 1979–82 when, in the southern hemisphere, it appears to be linked to the second branch by a bifurcation arm (Figure 8.2(b)). Although there appears to be a gap in the patterns between 1981 and 1983, they are re-established in 1984 (see also Figure 8.3(a)) and begin to follow the butterfly diagram for Cycle 22.

The shear decrease pattern, which represents the zones of flattening of the differential rotation curve, is shown by the dashed contours in Figure 8.2(b) and by solid lines in the lower panel of Figure 8.3(b). This pattern consists of relatively continuous zones, which begin near the poles at about sunspot

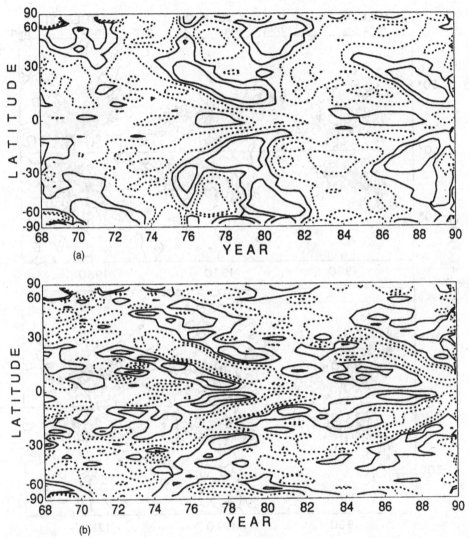

Fig. 8.2 (a) The torsional pattern in the rotation of the photospheric plasma (as determined by Doppler-shifts). The pattern is produced by subtracting an 11-year mean at each of 34 latitudes from an error-corrected rotation rate and averaging over 365 days. Contour intervals are -5, -2, 2, and 5 m s^{-1}, with negative contours (corresponding to flows at less than the local rotation rate) shown as dashed lines. (b) The torsional shear oscillation pattern derived from the centred difference between the torsional amplitudes in adjacent latitudes. Excess shear is indicated by the solid contours while reduced shear is represented by dashed contours. Contour intervals are -1.5, -0.6, 0.6, and 1.5 m s^{-1} deg^{-1}. (From Snodgrass 1991)

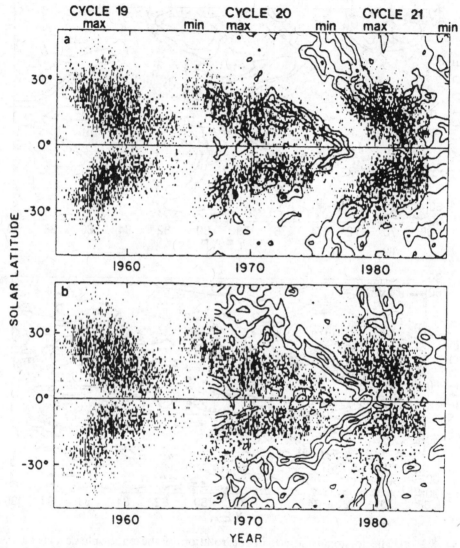

Fig. 8.3 (a) The excess shear pattern is compared with the butterfly diagram for Cycles 19–21. (b) The shear decrease pattern is compared with the butterfly diagram for Cycles 19–21. (From Snodgrass 1987)

minimum and drift steadily equatorwards, until they reach the equator at the maximum of the next cycle some 18 years later. The lower panel of Figure 8.3(b) shows that these zones lie in the gaps between adjacent wings of the butterfly diagram.

While emphasizing that perturbations in the surface velocity field as small as those associated with the torsional oscillations (~ 10 m s^{-1}, compared with the more random velocities of ~ 400 m s^{-1} associated with the granules and supergranules) cannot, in any sense, be said to cause or 'trigger' the surface eruptions of field, Snodgrass suggests that both the torsional and the magnetic activity patterns may be the surface shadows of some more potent phenomenon occurring within the convection zone.

Altrock (1988) has obtained recent data for the coronal emission at the NSO facility at Sacramento Peak. Daily observations of the corona have been made in the Fe XIV green line continuously since July 1975, with the exception of short gaps due to weather or equipment failure. The 40-cm-aperture coronagraph is used to form an occulted image of the corona, which passes through a narrow-band filter that spectrally discriminates at 100 kHz between the corona in the emission line and a continuum wavelength, a technique which allows the sky background to be subtracted electronically.

The entrance aperture of 1.1 arcmin diameter is scanned around the limb at several radii in order to detect local intensity maxima in the green-line emission as a function of latitude. At lower latitudes the coronal intensity maxima are known to occur over the main activity zones. Altrock infers that the occurrence of high-latitude emission maxima is related to the presence of high-latitude activity regions. Localized high-latitude intensity maxima are observed at 0.15 R$_\odot$ above the solar limb throughout the cycle. They evolve slowly over a period of days, consistent with the rotation over the limb of stable features, in a manner similar to the lower-latitude maxima that are connected with active regions. They persist long enough at a given latitude to be visible in long-term (e.g. annual) averages and, in Figure 8.4, averages over several rotations are used to derive the synoptic contour map for a period extending over 15 years from 1973 to 1988.

Two types of high-latitude activity zones can be identified in this diagram. A poleward-moving branch, which may be related to the poleward motions of the polar crown (see § 10.2), appeared near latitude 60° in 1978 (possibly earlier at lower latitudes). In 1979 an equatorward-moving region branched off from this poleward region (which continued on to reach the pole in 1980) at a latitude of 70° to 80°. It evolved approximately parallel to the main activity zone of the butterfly diagram, with the average latitude decreasing at a rate of roughly 5-6° per year. Near solar minimum, this branch evolved into the main activity zone of Cycle 22, and the emission continued its path towards the equator, where it should eventually disappear around 1998.

Altrock (1992) also reports that, as of early 1992, the pattern seen earlier appeared to be repeating itself. The poleward-moving branch became apparent near the beginning of 1988 near 60° latitude in the northern and southern hemispheres. The northern branch reached the pole during late

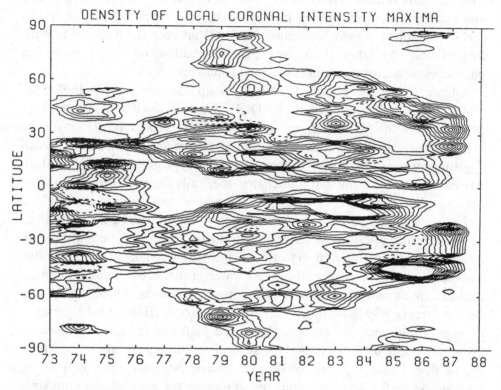

Fig. 8.4 Contours of annual averages of the number of intensity maxima are plotted as a function of latitude. The observations were made in Fe XIV 530.3 nm at a height of 0.15 solar radii above the limb with the Emission Line Coronal Photometer at the Sacramento Peak site of the National Solar Observatory in Sunspot, NM. (Courtesy R. C. Altrock, USAF Phillips Laboratory, Sunspot, NM.)

1989 to 1990, and polar emission effectively ceased at the end of 1990. The southern branch moved more slowly, and the southern-most emission regions reached the pole in mid-1991. South-polar emission was still occurring as of the latest observations. The equatorward-moving branch which may be the precursor of sunspot Cycle 23 became clearly established in the northern hemisphere near the beginning of 1990 at approximately 70° latitude. Currently the zone is at about 55°. The appearance of the equatorward-moving branch in the south was less dramatic but probably began in mid-1990 near 70°. Its position in 1992 was also near 55° latitude. The experience of previous cycles suggests that these equatorward-moving branches should continue steadily to the equator, where emission should cease about the year 2009.

In a different context, Legrand and Simon (1981) noted that enhanced geomagnetic activity related to the solar cycle arose from two components, one being the high-speed wind streams originating from high-latitude coronal holes and the other arising from sunspot activity. They pointed out that the coronal holes which give rise to these streams form shortly after the polar field reversals and last for 9–10 years. They further suggested that the formation of these high-latitude coronal holes may mark the beginning of the new cycle.

The geomagnetic activity generated by fluctuations in the solar wind particle streams is characterized by the geomagnetic indices *aa* and *Ap*. The *aa*-index measures the averages of the disturbances registered at two antipodal stations, while the *Ap*-index is the average of many disturbances distributed over the surface of the Earth. Thompson (1988) has analysed the structure of the geomagnetic disturbances through the cycle and found that the *Ap*-index exhibits two peaks, one approximately coincident with sunspot maximum, and the other occurring during the declining phase. He found that, while the amplitude of the first peak tends to correlate with that of the current sunspot cycle, the amplitude of the second tends to correlate with that of the following cycle, a result which is further discussed in Chapter 14. (Thompson's results are shown in Figure 14.1.)

Several different groups of scientists, working in very different areas, thus independently suggested that their data were related to a high-latitude component of solar activity, with which the large-scale torsional patterns were also involved, and proposed that this component is related to the new cycle, which begins at higher latitudes some 4–5 years before the appearance of the first sunspots of the new cycle, i.e. shortly after the maximum of the old one.

8.3 The onset of bipolar activity

K. Harvey (1992) has recently warned against an uncritical acceptance of the association of high-latitude activity with the following cycle, pointing out that the physical connections relating phenomena, such as coronal green-line emission, geomagnetic activity and torsional oscillations, and bipolar magnetic activity are not well understood. She has made a careful study of all bipolar magnetic regions across the full spectrum of sizes from large sunspot-producing active regions to small ephemeral regions and, by attempting to determine their membership of old or new cycles, has obtained estimates of the periods and therefore of the overlap of the bipolar magnetic cycles.

The orientation of the magnetic axes of active regions at mid-to-high latitudes is the primary criterion defining cycle membership, but neither latitude nor orientation provides an absolute indicator. Within the population of active regions, it has been established that a few per cent are of reversed polarity orientation for their cycle and hemisphere (e.g. by Richardson 1948, Smith and Howard 1968). Estimates range from 3.1% to 7.1% for regions larger than 315 millionths of a hemisphere (10^{-6} $H_\odot$). When such regions emerge at mid-latitudes within a few years before and after sunspot maximum, it seems reasonable to regard them as minor exceptions to the Hale–Nicholson law and associate them with the current cycle; but, when they occur towards the beginning or the end of a cycle, their allocation is less obvious.

Small active regions and ephemeral regions also show a preference for an orientation appropriate to their cycle and hemisphere. The latitude distribution of smaller regions relative to the mean latitude of the larger regions increases systematically with decreasing area from a poleward extent of $\sim 24°$ (i.e. $\sim 50°$ in heliographic latitude) for regions larger than 315×10^{-6} $H_\odot$, to $\sim 55°$ (i.e. to the poles) for regions with areas from 121 to 218×10^{-6} $H_\odot$. Further, for these small regions, the relative population of reverse polarity regions also increases with decreasing region size to 25% for smaller ephemeral regions with areas of $\sim 120 \times 10^{-6}$ $H_\odot$, such that the allocation of these regions to old or new cycles needs careful investigation.

Harvey *et al.* (1975) and Martin and Harvey (1979) have also found that the fraction of reversed polarity occurring in small and ephemeral active regions increases with poleward distance in the declining phase of the cycle. They reported that, in 1973, the preferred orientation of ephemeral regions in a latitude range from $\sim 30° - 60°$ in both hemispheres had reversed from that which had been observed in 1970 at these latitudes and from the preferred orientation of ephemeral regions at lower latitudes. These reversed-polarity ephemeral regions were offered as evidence of the occurrence of Cycle 21 activity three years before sunspot minimum; however, because of the small number of regions actually observed and the difficulties of identifying such regions at high latitudes, the statistical significance of the result was in some doubt.

A reversal of the preferred orientation in the population of high-latitude ephemeral regions (ERs) was also observed ten years later, during the decline of Cycle 21 (Wilson *et al.* 1988), and was subjected to a more careful scrutiny. The histograms in Figure 8.5 show the distributions of ERs as a function of latitude for the years 1980–86. The unhatched histograms represent the total number of ERs in a given sine latitude zone; the hatched histograms represent the ratio

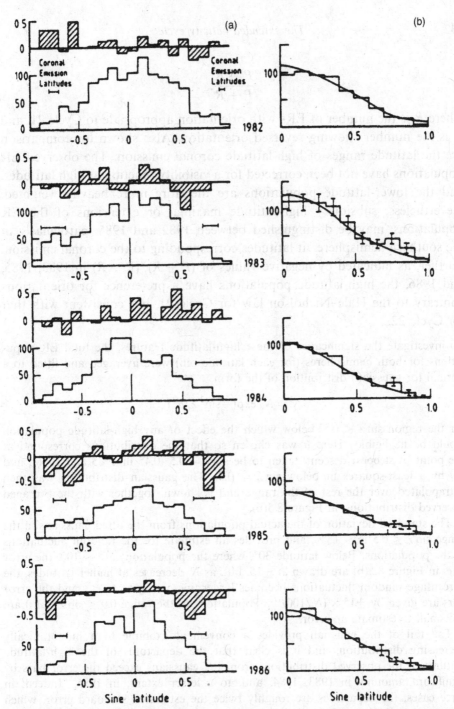

Fig. 8.5 (a) The unhatched histograms represent the total ER populations, $P + R$, plotted against sine-latitude for the years 1982–1986. The hatched histograms show the relative excess of normally oriented ERs for Cycle 21, $(P - R)/(P + R)$. Negative values therefore indicate a preference for orientations appropriate for Cycle 22. The horizontal bars indicate the high-latitude coronal emission. (b) The histograms show the ER populations for the sine-latitude ranges, averaged over both hemispheres. The error bars are shown for all latitude ranges for 1983 and at selected latitudes in other years. The curve represents a gaussian distribution fitted to the low-latitude values. (From Wilson *et al.* 1988)

$$\frac{P - R}{P + R},$$

where P is the number of ERs with orientation appropriate to Cycle 21, and R is the number showing reversed orientation. Also shown for comparison are the latitude ranges of high-latitude coronal emission. The observed ER populations have not been corrected for a visibility function at high latitudes, and the lower-latitude populations are therefore more heavily weighted. Nevertheless, subsidiary high-latitude maxima, or extensions of the ER populations, may be distinguished between 1982 and 1985, particularly in the southern hemisphere, at latitudes corresponding to the coronal emission. Further, as indicated by negative values of $(P - R)/(P + R)$ in 1983, 1985, and 1986, the high-latitude populations have a preference for orientations contrary to the Hale–Nicholson law for Cycle 21, but consistent with that for Cycle 22.

To investigate the significance of these high-latitude features, the total ER populations for both hemispheres (for each latitude bin) were averaged and fitted to a normal (or gaussian) distribution of the form

$$N = N_0 \exp(-\sin^2 \lambda / \sigma^2)$$

for the region $\sin \lambda \leq 0.5$, below which the effect of any high-latitude population would be negligible. Here σ was chosen so that the distributions correspond at the point of steepest descent, taken to be $\sin \lambda = 0.5$ (0.45 in 1985 and 1986), and N_0 by a least-squares fit below $\sin \lambda = 0.5$. The gaussian distribution was then extrapolated over the rest of the range and is shown, together with the averaged observed distributions, in Figure 8.5(b).

The standard deviation of the actual populations from the fitted gaussians in the range $\sin \lambda \leq 0.5$ is $\sim 15\%$, and provides an estimate for the random fluctuations in the populations. Below latitude $30°$, where the populations are ~ 100, the error bars in Figure 8.5(b) are drawn at ± 15, but, as N decreases at higher latitudes, the percentage random fluctuations becomes larger, increasing as $N^{-1/2}$, and the error bars are given by $\pm 15 \times (N/100)^{1/2}$. Populations in the range $0.9 \leq \sin \lambda \leq 1.0$ are too small to estimate an error.

The tail of the gaussian provides a convenient example of a monotonically decreasing distribution, and it is clear that the departures of the high-latitude features of the observed distributions from the gaussians exceed the error bars by significant amounts in 1983, 1984, and, to a lesser extent, in 1985. Indeed, in three cases, the departures are roughly twice the estimated standard error, which represents a 90% probability that the departures are real.

The evidence for the existence of a distinct high-latitude ER population emerging in 1983, three years before sunspot minimum in 1986, is therefore significant. Figure 8.5(b) shows no significant high-latitude population in

1982, but Figure 8.5(a) indicates that somewhat smaller subsidiary maxima occur in different latitude bins in opposite hemispheres and are concealed by the averaging process in Figure 8.5(b).

Because these high latitude regions also showed a clear preference for reversed, rather than normal, orientations in 1983, 1985, and 1986 (see Figure 8.5(a)), these results may be interpreted as evidence for the occurrence of new-cycle high-latitude bipolar activity in 1982–83, i.e. at least three years before the appearance of the first identified sunspots of Cycle 22 in 1986.

8.4 The end of bipolar activity

As discussed in §9.7 below, there are difficulties associated with the assignment of low-latitude regions to the old or the new cycle: the frequency of occurrence of reverse-polarity regions increases during the minimum phase, and it is not immediately obvious whether such regions are new-cycle regions emerging at unusually low latitudes or old-cycle regions from the opposite hemisphere which have drifted across the equator. However, the fact that they are spatially interspersed with the more prevalent properly oriented regions suggests to Harvey (1992) that these regions should be associated with the old cycle rather than the new. Region 736, which emerged in CR 1778 and is shown in Figure 9.6, is a good example of such a region.

Consistent with this interpretation is the fact that the number of regions within 7° of the equator does not begin to increase until 3 years after minimum, which indicates that regions of the incoming cycle do not reach the equatorial zone until well into the cycle. Harvey determined the equatorward boundary of the regions of the incoming cycle and found that it reached the equator 3 to 5 years after sunspot minimum. She considered active regions occurring between this boundary and the equator to be old-cycle, regardless of their polarity orientation.

Using these criteria, she determined that the period between first and last magnetic bipoles of Cycles 14–21 lay in the range 13.2–14.8 years. A further study by Lean (private communication) of the butterfly diagram constructed from calcium plage data has extended the lifetimes of Cycles 19 and 20 to 14.0 and 15.7 years respectively. Harvey's butterfly diagram of sunspot regions, combined with indicators of the latitudinal extent of calcium plage regions and ephemeral regions, is reproduced in Figure 8.6. The wings of the butterfly diagram are here *extended* back, so that there is an unambiguous overlap between adjacent cycles.

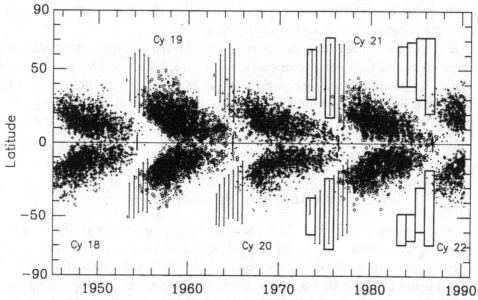

Fig. 8.6 The butterfly diagram of sunspot regions from 1945 to 1991, together with the latitudinal extent of CaII regions, indicated by vertical lines, and ephemeral regions identified with the new cycle, indicated by boxes. The times of sunspot minima are indicated by vertical lines on the equatorial axis. (From Harvey 1992).

8.5 The poleward branch

Considerable attention has recently been paid to the poleward arm of the pattern, which can be discerned in the synoptic diagrams representing the torsional shear and the coronal density maxima (Figures 8.2, 8.3, and 8.4). The poleward arm is particularly obvious in the southern hemisphere in each diagram, where, in 1976 (i.e. at sunspot minimum), the patterns bifurcate at latitudes of $40° - 50°$, one branching towards the pole, which it reaches in 1980–81, and the other following the butterfly diagram towards the equator. A similar, but weaker, poleward branch can be seen in the northern hemisphere.

These poleward branches are consistent with observations of McIntosh (1992), who noted that the polar crowns of Hα filaments are observed at latitudes $40° - 50°$ near sunspot minimum and progress polewards during the rising phase of the cycle, their arrival at the poles coinciding with the polar field reversals (see § 10.2, Figure 10.1).

Sivaraman and Makarov (1992) have studied the evolution of high-latitude faculae and find a rather different pattern in which the polar faculae make their first appearance in the latitude zone $40° - 50°$, shortly after the polar field reversals, i.e. near sunspot maximum. The zones of emergence migrate slowly polewards, eventually reaching latitudes of 70° − 80° as the cycle progresses. Their poleward progress is therefore some 4–5 years out of phase with the torsional shear and the coronal density maxima patterns.

Obridko and Gaziev (1992) describe a zone of unipolar magnetic field which arises at sunspot maximum at equatorial latitudes and migrates polewards over a period of 17 years, with an average width of 11 years. This poleward migration is also consistent with studies of sunspot motions described by Ribes (1986), who interpreted these as 'roulaids', or convective rolls, which progressed polewards in company with the magnetic features.

Bortzov, Makarov, and Mikhailutsa (1992) find that a new cycle of coronal activity, having two components in each hemisphere, commences after the polar field reversal. The first component is identified with the polar faculae which appear at latitude 40° and migrate polewards; the second component shows up at latitude 40° five or six years later. Bortzov *et al.* therefore infer that the global coronal activity has a duration of 16–17 years.

The reader may find these apparently contradictory results regarding the time and place of origin of the poleward branch and its duration somewhat confusing. Some clarification has, however, been achieved in a more sophisticated approach by Stenflo (1992), who has obtained the radial component of the axisymmetric magnetic field $B_r(t, \theta)$ at colatitude θ for each rotation by averaging the synoptic magnetic field maps over longitude. $B_r(t, \theta)$ is then expanded in spherical harmonics:

$$B_r(t, \theta) = \text{Re} \sum_{\ell=1}^{N} a_\ell e^{2\pi i \nu_\ell t} P_\ell(\cos \theta), \tag{8.1}$$

where $P_\ell(x)$ is the Legendre polynomial of degree ℓ. Stenflo limits the sum to the first seven odd modes and determines the coefficients by a least-squares fit to the axi-symmetric magnetic data, thus deriving what he calls a 'modally clean butterfly diagram' (Figure 8.7).

The magnetic pattern which constitutes a 'cycle' consists of several 'components'. It first appears at latitudes of 40° − 50° near minimum (e.g. 1964); one branch migrates equatorwards over ∼ 22 years, overlapping with the butterfly diagram of the *next-but-one* cycle. A broad poleward branch also appears at the beginning of the cycle, first reaching the pole near sunspot maximum (corresponding to the polar reversal), achieving its greatest polar intensity at the next minimum, and extending on to the next polar reversal. This poleward feature is consistent with the account of Obridko and Gaziev (1992) but begins at higher latitudes. The other accounts may be understood

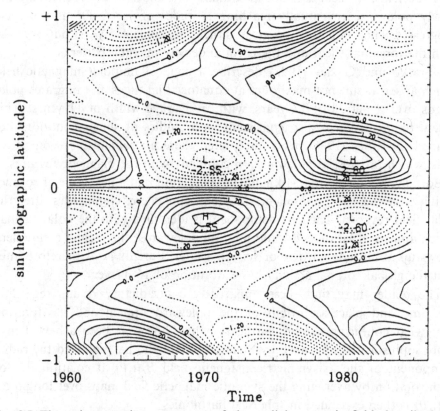

Fig. 8.7 The axisymmetric component of the radial magnetic field described as a superposition of seven discrete odd-parity modes, $l = 1, 3, \ldots, 13$, each varying sinusoidally with a period of 22 years. (From Stenflo 1992)

by associating the poleward progression of the faculae with the 'ridge' of the magnetic feature, and the green line emission, the torsional shear, and the polar crown progression with the neutral lines of this poleward branch. It is interesting to note from the contour spacings that, whereas the non-axisymmetric phenomena (e.g. sunspots) are much stronger at lower latitudes than at the poles, the axisymmetric components are of approximately equal strength.

It should be remembered that the azimuthally averaged fields obtained by Stenflo describe the net poloidal components of the surface fields at a given latitude, the toroidal components being eliminated by the averaging process.

8.6 Discussion

The activity phenomena described above range from small magnetic bipoles (ERs) through green-line coronal emission maxima to geomagnetic disturbances, yet although they emerge during one sunspot cycle (as defined by consecutive minima) they all appear to be related (by latitude, orientation, or both) to the *following* cycle. If this association has a physical basis, the lifetime of an activity cycle extends beyond the 11-year period of the traditional sunspot cycle and the term *extended activity cycle* (EAC) has been introduced to describe this association. We now consider the implications of this association for an understanding of the cyclic phenomenon.

The EAC does *not* imply an extension of the period between successive maxima (or minima) of the cycle. Instead, it challenges the one-dimensional picture in which all activity phenomena which appear between successive minima are deemed to belong to a particular cycle, while phenomena which occur before the first minimum or after the second belong to the adjacent cycles.

In its place, the EAC offers an essentially two-dimensional picture, in which activity phenomena may be associated with a particular cycle according to the latitude and epoch of their emergence and, in the case of magnetic bipoles, according to the orientation of their magnetic axes. As a result, the wings of the butterfly diagram may be extended to higher latitudes, so that they overlap and form a 'herringbone' pattern. This conclusion is drawn specifically in the work of many of the authors cited above (e.g. Altrock 1988, Snodgrass 1987). At the 1991 Solar Cycle Workshop in Sunspot, New Mexico, it appeared that many regarded the conclusion as self-evident; some, including Stenflo (1992), considered that the concept was 'superfluous'. Various somewhat irreverent attitudes to the EAC are summarized in the cartoon shown in Figure 8.8 (for which the author disclaims any responsibility).

The EAC concept is, however, not universally accepted. Harvey has emphasized that the relation between coronal emission and surface bipolar magnetic activity is unclear and that the torsional oscillations are not activity phenomena at all. On several occasions, Gilman (e.g. Gilman, Morrow, and DeLuca 1989) has re-asserted the one-dimensional approach, in which both the high-latitude and low-latitude phenomena appearing between successive minima must be associated with the cycle defined by those minima. He argues that there is no reason why the magnetic orientation of high-latitude bipolar phenomena of a given cycle must be the same as that of the low-latitude phenomena.

THE XTENDED CYCLE

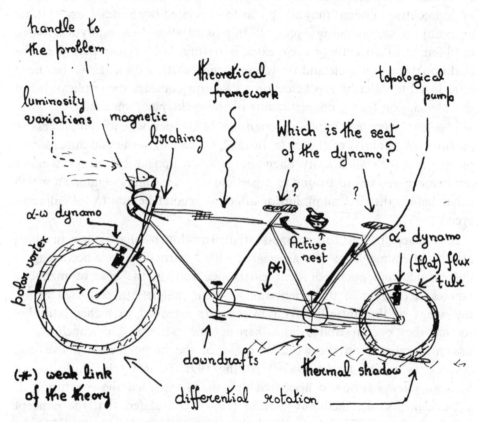

Fig. 8.8 A diagram of the *Xtended Cycle* constructed at a party held during the Sunspot meeting of the Solar Cycle Workshop in 1991. The author disclaims any responsibility but understands that Jean-Paul Zahn is liable for the drawing, Sydney D'Silva for the lettering, and various irreverent (and possibly inebriated) astrophysicists for the captions.

These different viewpoints serve only to highlight the lack of any established physical understanding of the sub-surface processes which generate the observed surface phenomena. Relaxation models, such as Babcock's, require a 'relaxation' interval between successive cycles, in which the global poloidal field is dominant. These models require that the first bipolar regions of the new cycle can emerge only after this poloidal field has undergone some degree of winding by the differential rotation (see §7.2). If Gilman's view is correct, such models must remain in consideration; but, if bipolar regions of a given cycle are present for at least 15–16 years, such models face severe difficulties (see § 7.3). If the new cycle first appears at high latitudes as part of the polar field reversals, then relaxation models may be rejected.

On the basis of the EAC data, Parker's (1955) simple dynamo wave model (see § 7.8), in which two sub-surface toroidal bands of magnetic field co-exist within each hemisphere and propagate equatorwards, seems to be the more appropriate. The drawbacks of this model, however, are associated with the direction of propagation of the waves and the presence of the poleward-migrating branches.

Gilman, Morrow, and DeLuca (1989) have proposed a version of the dynamo wave model in which the wave begins at mid-latitudes at sunspot minimum and consists of two components, one of which propagates equatorwards and generates the sunspot cycle, while the other propagates polewards, giving rise to the phenomena of the poleward branch described above. According to Gilman *et al.*, the high-latitude bipoles emerging during the declining phase of the old cycle should be regarded as part of the poleward component of the current cycle, rather than as a precursor of the next. This model is further discussed below (§12.5) in relation to helioseismology data.

8.7 Conclusion

It is not possible, at this stage, to make a clear choice between these two interpretations of the data. In common with many others, however, the author finds the EAC concept compelling. There can be little doubt that the sunspots of a given cycle are present for periods of ∼ 14 years and that the high-latitude spots with new-cycle orientations overlap the low-latitude spots of the old by ∼ 3 years. It is significant that no-one has suggested that these spots should be regarded as members of the old cycle and, in view of this, there can be no *a priori* reason why other phenomena of activity appearing at high latitudes at and before this time should not also be regarded as part of the new cycle.

The results of Martin and Harvey demonstrate the presence of a high-latitude component of ephemeral active regions and larger magnetic bipoles without sunspots during the declining phase of the old cycle. Although the orientations of these regions have a wider scatter, there appears to be a distinct preference for orientations corresponding to the next, rather than the current, cycle. Without a detailed model to account for the emergence of these small regions, their allocation to one or other cycle is perhaps a matter of definition, but, consistent with the identification of some pre-minimum spotted active regions of appropriate latitudes and orientations with the new cycle, it seems not unreasonable to make the same allocation for the smaller magnetic bipoles. (For a plausible model which accounts for the emergence of these small magnetic bipoles, see §9.9 below).

The evidence provided by the torsional shear patterns, by the coronal green-line emission, and by the correlation between the geomagnetic disturbance indices late in the old cycle and the amplitude of the maximum of the new cycle is less direct than that provided by the magnetic bipoles. These data nevertheless provide considerable independent support for the view that phenomena associated with the new cycle do make their appearance at high latitudes during the post-maximum phase of the old cycle.

Stenflo's analysis of the poloidal magnetic fields, which shows a pattern extending over almost two 11-year cycles and includes both poleward and equatorward branches, supports and extends the suggestion that new-cycle phenomena appear during the declining phase of the old cycle. It would seem that the phenomena associated with an extended activity cycle may first appear at mid-latitudes near sunspot minimum; shortly thereafter the pattern bifurcates into two branches, a broad poleward branch for which the neutral line progresses rapidly polewards and is associated with the polar field reversal near sunspot maximum of the *next* cycle, and an equatorward branch which merges with the butterfly wing of the *next-but-one* cycle.

Since the two branches have a common bifurcation point, it is reasonable to infer a common physical origin. Stenflo (1992) has pointed out that each harmonic of his expansion contributes to *both* the polar and the equatorial branches. He argues that such patterns are not uncommon in dynamo theory and that an approach which is more sophisticated than the simple dynamo wave model is necessary. Before this can be attempted, however, results of studies of the large-scale fields, the polar field reversals, helioseismology, and dynamo theory must be reviewed.

References

Altrock, R. C.: 1988, in *Solar and Stellar Coronal Structure and Dynamics.*, ed. R. C. Altrock, *NSO/Sacramento Peak Summer Workshop*, Sunspot, New Mexico, 414.

Altrock, R. C.: 1992, *Bull. Am. Astron. Soc.*, **24**, 746.

Babcock, D. D.: 1961, *Astrophys. J.*, **133**, 572.

Bortzov, V. V., Makarov, V. I., and Mikhailutsa, V. P.: 1992, *Solar Phys.*, **137**, 395.

Bretz, M. C., and Billings, D. E.: 1959, *Astrophys. J.*, **129**, 134.

Dodson, H. W., Hedeman, E. R., and Mohler, O. C.: 1982, *World Data Center A*, Report UAG-83, 1.

Gilman, P. A., Morrow, C. A., and DeLuca, E. E.: 1989, *Astrophys. J.*, **338**, 528.

Giovanelli, R.: 1964, *Observatory*, **84**, 57.

Harvey, K. L., Harvey, J. W., and Martin, S. F.: 1975, *Solar Phys.*, **40**, 87.

Harvey, K. L.: 1992, in *Proceedings of the National Solar Observatory/Sacramento Peak 12th Summer Workshop*, ed. K. L. Harvey, San Francisco, California, 335.

LaBonte, B. J., and Howard, R.: 1982, *Solar Phys.*, **75**, 161.

Legrand, J. P., and Simon, P. A.: 1981, *Solar Phys.*, **70**, 173.

Leroy, J. L., and Noens, J. C.: 1983, *Astron. and Astrophys.*, **120**, L1.

McIntosh, P. S.: 1992, in *Proceedings of the National Solar Observatory/Sacramen-to Peak 12th Summer Workshop*, ed. K. L. Harvey, San Francisco, California, 14.

Martin, S. F., and Harvey, K. L.: 1979, *Solar Phys.*, **64**, 93.

Obridko, V., and Gaziev, G.: 1992, in *Proceedings of the National Solar Observatory/Sacramento Peak 12th Summer Workshop*, ed. K. L. Harvey, San Francisco, California, 410.

Parker, E. N.: 1955, *Astrophys. J.*, **122**, 293.

Ribes, E.: 1986, *C. R. Acad. Sci. Ser. Gen. Vie Sci.*, **3**, 305–25.

Richardson, R. S.: 1948, *Astrophys. J.*, **107**, 78.

Sivaraman, K. R., and Makarov, V. I.: 1992, in *Proceedings of the National Solar Observatory/Sacramento Peak 12th Summer Workshop*, ed. K. L. Harvey, San Francisco, California, 415.

Smith, S. F., and Howard, R.: 1968, in *Structure and Development of Solar Active Regions*, ed. K. O. Kiepenheuer, D. Reidel Publ. Co., Dordrecht, Holland, 33.

Snodgrass, H. B.: 1987, *Solar Phys.*, **110**, 35.

Snodgrass, H. B.: 1991, *Astrophys. J.*, **383**, L85.

Stenflo, J. O.: 1992, in *Proceedings of the National Solar Observatory/Sacramento Peak 12th Summer Workshop*, ed. K. L. Harvey, San Francisco, California, 421.

Thompson, R. J.: 1988, *Australian J. Phys.*, **25**, 17.

Trellis, M.: 1957, *Ann. d'Astrophys. Suppl.*, **5**.

Waldmeier, M.: 1957, *Die Sonnenkorona*. Vol. II., Birkhäuser, Basel.

Wilson, P. R., Altrock, R. C., Harvey, K. L., Martin, S. F., and Snodgrass, H. B.: 1988, *Nature*, **333**, 748.

9

The origin of the large-scale fields

There's some who'll say that what I've said is wrong,
while others claim they've known it all along.
T. Simon, introducing the 7th Cool Stars Workshop

9.1 Introduction

Although not the most spectacular phenomenon of the solar cycle and undetectable in stellar cycles, the large-scale magnetic field patterns on the Sun play an important role in solar cyclic activity and in the attempts to understand the solar cycle discussed in Chapter 6. In relaxation models such as the Babcock model (see § 6.2), the reversal of the polar fields by the poleward drift of large-scale fields is crucial, and in the Leighton–Sheeley flux-transport model the large-scale fields arise solely from the decay of active-region fields.

However, magnetic flux emerges at the surface of the Sun in the form of magnetic bipoles whose dimensions range across a wide spectrum, from the largest active regions with dimensions up to 100 000 km, down to the small intra-network elements of order 500 km. Stenflo (1992) has noted that the *total* flux emerging in the smaller elements exceeds that emerging as large active regions by several orders of magnitude, and there is no *a priori* reason why regions from the large-scale end of this spectrum should be the only contributors to the large-scale field patterns and thus to the reversal of the polar fields.

In this chapter we describe recent studies of the evolution of the large-scale field patterns at the beginnings of Cycles 20, 21, and 22. In Chapter 10 we look at the polar fields near the maximum of Cycle 22.

9.2 Observing the large-scale magnetic fields

The measurement of solar magnetic fields depends on the splitting of magnetically sensitive spectral lines in the presence of a magnetic field, which is known as the *Zeeman Effect*. The flux density, or induction, of the magnetic field component along the line of sight is measured in a *magnetograph* by calibrating the wavelength displacement of the spectral line when the circular polarization passed by the analyser is reversed. The 'Zeeman splitting' so measured is, however, an average of the Zeeman splittings from all the atoms within a small volume or 'surface slice' of the photosphere. This surface slice has a thickness equal to the narrow range of photospheric depths where the line formation takes place, and a cross-section determined by the resolution of the magnetograph.

Away from active regions, the magnetic field lines threading the photosphere are not uniformly distributed but are concentrated into thin, sub-arcsecond diameter flux tubes (Howard and Stenflo 1972), and the fraction of the surface traversed by these intense tubes in a given region is called the *filling factor*. Further, fields of both polarities may be present in any such surface slice, so that a magnetograph measurement is an average in which the strong fields within the tubes are diluted both by the intervening field-free regions and by cancellations with opposite polarity tubes (e.g. bipoles) traversing the slice. This volume average is also affected by the angle, θ, between the field lines and the line-of-sight. Away from active regions, the field lines traversing the photosphere are nearly radial (Howard and LaBonte 1981), although there is slight systematic east-west tilt (Howard 1974). Near disk-centre the line-of-sight intensity of fields inclined to the radial direction is decreased by the factor $\cos \theta$, but this is offset by the increase in the path length of the field line through the surface slice (which is increased by $\sec \theta$). For observations made near disk-centre, therefore, the magnetograph measures a mean flux density which is not affected by the angle of the field.

Away from disk-centre the weakening of the line-of-sight component of the radial field lines is partly compensated by the increased area of the photosphere subtended by the entrance window, provided the filling factor and the polarity distribution are uniform. Tubes inclined to the radial direction, however, create a problem for measurements of fields near the limb. In the case of tubes inclined towards disk-centre, both the reduced angle with the line-of-sight and the increased path length through the slice artificially enhance their contribution to the magnetograph signal and, owing to the complexity of the geometry of the magnetic structures, this may not be balanced by the contribution from tubes inclined away from disk-centre.

Thus the characteristic east-west tilt of 10° creates a systematic error whereby both positive and negative fields appear enhanced in the eastern hemisphere compared to those in the western hemisphere.

Various technical difficulties also contribute to the uncertainties of magnetograph measurements. Among these are: (1) polarization produced in the optics of the telescope and the spectograph, which can create a bias that depends on the angle of the Sun in the sky; (2) imperfect cycling of the optically active crystals which are used to analyse the polarization; and (3) saturation of the magnetograph by strong, concentrated magnetic fields. For an instrument such as the 150-foot tower at Mount Wilson, saturation occurs in the $\lambda5250.2$ Fe I line when the field strength averaged over the surface slice approaches 100 G. This results from the high g-factor of the line and the geometry of the exit slit. On the other hand, the magnetograph is a sensitive instrument, capable of recording fields as small as ~ 0.25 G.

The resolution of the magnetograph is the mean linear dimension of the solar surface area subtended by the entrance slit at disk centre. A *magnetogram* is a map of the magnetic field over an extended area obtained by moving the entrance window systematically across an image of the solar surface, examples of which are shown in Figures 9.1 and 9.4. A *unipolar* region is one where contiguous measurements over an extended area register the same polarity. Regions appear as unipolar in magnetograms only when the sizes of the individual magnetic structures are much smaller than the resolution of the instrument.

Magnetograms able to resolve individual flux tubes show that there are many elements of opposite polarity present throughout such regions, so that, although the average field strength is of the polarity displayed at low resolution, the high-resolution magnetogram exhibits a 'pepper and salt' pattern of small-scale features of positive and negative polarities. These individual features have linear dimensions of ~ 1 arcsec or less, while the Mount Wilson magnetograph has a resolution of ~ 10 arcsec, which makes it well suited to observations of the large-scale or unipolar regions, although it is not capable of resolving the individual fine structures.

We are here concerned with the evolution of the global patterns of the surface fields, as revealed by regular synoptic observations. However, as we shall see, high-resolution observations of small-scale fields are also of considerable importance to an understanding of the nature and dynamics of this evolution.

Many observatories contribute to the international synoptic magnetic data-base. Full-disk daily magnetograms are obtained by the National Solar Observatory's Kitt Peak Station (NSO), the Wilcox Solar Observatory at Palo Alto, and the Mount Wilson Observatory (MWO). These (and many other data) are available in the *Solar-Geophysical Data Prompt Reports* (SGD), published by the NOAO National Geophysical Data Center, Boulder, Colorado. The Big Bear Solar Observatory of the California Institute of Technology also contributes to the data base of both high- and low-resolution magnetic field observations.

From these daily records, synoptic data grids, or charts, which represent the distribution of surface magnetic phenomena during one complete Carrington rotation (CR, see §2.5) are prepared. Each grid represents 360° of Carrington longitude and 180° of latitude and is constructed using an algorithm, whereby the data from a daily magnetogram are mapped onto the grid, the central meridian for each day being identified with a particular Carrington longitude (CL). A typical example of an NSO synoptic picture for a complete solar rotation, that for CR 1787 (April 1987), is reproduced in Figure 9.1. The display consists of two panels: the upper panel represents the relative line-of-sight intensity of the magnetic fields; the lower panel shows the sign of the net polarity of the field per pixel (black is negative). The upper panel thus displays the location and size of the active regions, and the lower the large-scale field patterns.

The time scale for magnetic activity features is frequently less than one solar rotation, and the algorithm from which the maps are constructed places a greater weight on the central meridian configuration of a feature than on that near the limb. Further, variations occurring on the invisible hemisphere are, necessarily, neglected altogether. Features which emerge near the east limb should therefore be well-represented on these charts; conversely, those which emerge near the west limb and reach maturity on the invisible hemisphere are less accurately portrayed.

The lifetimes of moderate-to-large sunspots generally exceed one rotation and may extend over three or four, and so, for these features, accurate representation on synoptic charts is not a serious problem. The lifetimes of regions which produce only one or two small spots are, however, considerably shorter and may be of the order of 13 days, i.e. half a solar rotation. While synoptic charts provide an important means of studying the rotation-by-rotation evolution of the large-scale fields, they should be studied in conjunction with the corresponding daily magnetograms in order to determine the phase of any smaller, short-lived features and the reliability of their representation on the charts.

The pixel size of the NSO charts is ~ 1°, and, because of the 'pepper and salt' nature of the field patterns seen at this resolution, it is not always easy to identify the large-scale regions on this display. The somewhat lower

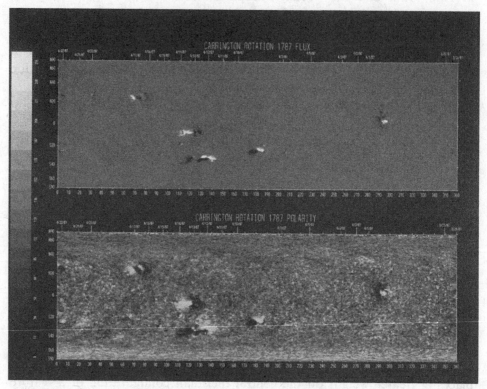

Fig. 9.1 Synoptic charts of magnetograms from the Kitt Peak station of the National Solar Observatory for Carrington rotation 1787.

resolution synoptic data from Mount Wilson are provided in numerical form on a 4° by 4° two-dimensional grid, and it is easier to construct synoptic contour maps from these. The map regions derived from the MWO data for CR 1787 are shown in Figure 9.2(a). Comparison with Figure 9.1 (which shows the NSO data for the same rotation) illustrates the advantages and defects of these different presentations.

Yet another method of displaying magnetic-field data is the Hα synoptic chart, shown for the same rotation in Figure 9.2(b). Here the positions of the neutral lines of the large-scale fields are inferred from observations of the Hα filaments (see McIntosh 1972a, 1972b) and from patterns of fibrils arranged to form filament channels. These structures permit the mapping of continuous neutral lines interconnecting the large-scale and the active-region fields, and it is assumed that the unipolar regions of the large-scale fields are located between these neutral lines. A comparison between Figures 9.2(a)

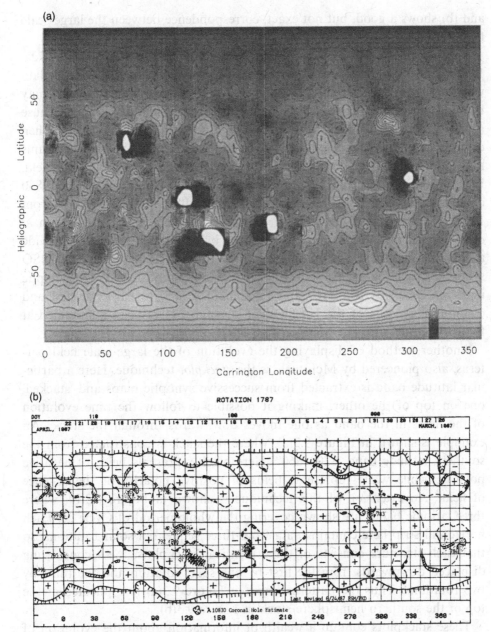

Fig. 9.2 (a) Synoptic contour map of magnetic fields derived from data provided by the Mount Wilson Observatory for Carrington rotation (CR) 1787. (b) Hα synoptic chart of magnetic fields for CR 1787 (26 March to 23 April 1987). Lines of polarity inversion (neutral lines) are mapped from filaments (cross-hatched areas on solid lines), filament channels (solid lines alone), and plage corridors (solid lines through active regions (dots)). Dashed lines are estimated continuations of neutral lines based on patterns of polarity and continuity with patterns on earlier solar rotations. Coronal holes are outlined with solid lines with tick marks directed toward the centre of the hole. The numbers near active regions are the last three digits of the serial numbers assigned by NOAA Space Environment Services Center.

and (b) shows a good, but not exact, correspondence between the large-scale field patterns.

The use of the Hα synoptic chart was pioneered by McIntosh (1972a, 1972b) and provides a valuable complement to direct magnetograph data when the latter are available. The locations of the inferred neutral lines may be more precise than those derived from line-of-sight magnetograms because (i) perspective difficulties may introduce a false neutral-line effect when diverging fields are observed away from disk-centre, and (ii) magnetograms may show only a broad, vague neutral zone between regions of weak field. In addition, the locations of the coronal holes, as indicated by He 10830 data, have been included in the charts since 1980, and 10830 data from 1975–79 are included in the most recent atlas of Hα charts (McIntosh *et al.* 1991). Further, the Hα charts produced by McIntosh since 1964 provide a valuable data-base extending over three solar cycles, whereas the NSO synoptic magnetogram maps have been available only since 1977. Synoptic maps produced by Makarov and Sivaraman extend to earlier cycles, and continuous records of the Hα filament structures have been maintained at the Paris Observatory at Meudon since 1920.

Another method of displaying the evolution of the large-scale field patterns, also pioneered by McIntosh, is the *stackplot* technique. Here a particular latitude band is extracted from successive synoptic maps and 'stacked' one on top of the other, making it possible to follow the time evolution of the fields in the band as one moves downwards through the stack. An example is shown in Figure 9.3(a), where the stack is constructed from the southern latitude band, S20°–S60°. Here black represents fields of negative polarity, while white represents fields of positive polarity or the absence of field. An extension of this technique is shown in Figure 9.3(b), where three latitude bands are shown covering the whole period of Cycle 20. Each horizontal strip here contains data from three sequential solar rotations, so that each structure consists of three identical stackplots, the second being displaced downwards from the first by one solar rotation and the third by two rotations. Thus the pattern shown in Figure 9.3(a) is triplicated at the top of the southern hemisphere stack in Figure 9.3(b).

These stackplots contain a wealth of information about the evolution of the large-scale field patterns in compact form. Because features at higher latitudes rotate more slowly than the mean Carrington rate from which these maps are constructed, high-latitude features are displaced backwards in longitude (i.e. to the left) in sequential Carrington rotations and are represented as streaks running downwards and to the left of the stackplot. In a single stackplot, these streaks eventually suffer 'wrap around' and

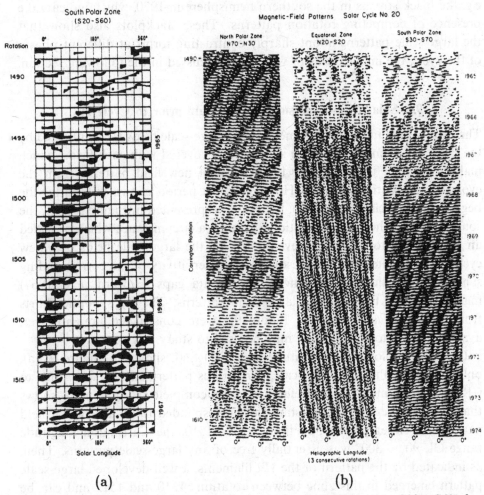

Fig. 9.3 (a) A stackplot of Hα synoptic charts for the latitude zone S20°−S60° for CRs 1486–1520 is shown. (b) A stackplot of Hα synoptic charts for polar and equatorial latitudes for Cycle 20. Note that each horizontal strip contains data for three consecutive rotations.

reappear at the right-hand end of the strip. However, by placing data from three solar rotations side by side, as in Figure 9.3(b), the drift of high-latitude features with respect to the standard Carrington longitude may be followed for several years without interruption from the (arbitrary) choice of zero Carrington longitude. In following these patterns, it should be remembered that, for example, the three white diagonal streaks, which are very obvious in the southern hemisphere between 1966 and 1968, represent the evolution of only a single feature on the Sun.

In some regions, streak patterns with two distinct inclinations can be seen, e.g. the black streaks in the southern hemisphere in 1970, which indicate the presence of competing rotation patterns. These stackplots also show that the large-scale patterns adopt sharply contrasting forms at different phases of the cycle, the implications of which are discussed in the following section.

9.3 Solar fields near sunspot minimum prior to Cycle 20

The study of the initial development of the large-scale fields is complicated by the fact that, at most times, the solar disk is covered with such fields, which make it difficult to distinguish the growth of new field patterns from the evolution of existing patterns. However, for a period of two years after the beginning of Cycle 20 (1964–66, the period represented in Figure 9.3(a)), the southern hemisphere was exceptionally quiet, a circumstance which provided an opportunity to study the development of the large-scale fields of a new cycle. Unfortunately, the low resolution and sensitivity of the corresponding synoptic magnetic data and the frequent data gaps (Howard *et al.* 1967) made it difficult to detect the large-scale patterns. The Hα synoptic charts from which the stackplots of Figure 9.3 were constructed are, however, available (McIntosh 1979) and may be used to study this period.

In the stackplot for the latitude zone S20–S60, shown in Figure 9.3(a), only a single, coherent, large-scale, continuous pattern can be seen prior to CR 1512 (September 1966). Reference to the complete synoptic maps shows that the white regions in this stackplot corresponded to an absence of field patterns of either polarity, so that, in CR 1510, the Carrington longitude range CL 90° − 300° was essentially free of any large-scale patterns. Then, as indicated by the pattern of the Hα filaments, a well-developed large-scale pattern emerged in this zone between rotations 1510 and 1512 and can be followed through the stackplots in Figure 9.3(b) for almost two years. The *Cartes Synoptiques*, published by the Meudon Group in France, confirm the sudden appearance of these filaments in CR 1512 and their subsequent development as described by McIntosh (1981).

Between June 1965 and September 1966, no significant sunspot groups emerged in the southern hemisphere, and the few active regions that formed in the longitude range 90° − 300° prior to the establishment of the second large-scale pattern in CR 1512 were small and short-lived. Indeed, it was not until rotation 1515 that the first large sunspot group of the hemisphere and cycle emerged on the most distinctive neutral line of the large-scale pattern which had formed three rotations earlier. This group contributed to the large-scale field but did not significantly disturb the pattern. Although other active regions continued to form, this neutral line and the pattern associated with it remained distinguishable for a further two years. Such behaviour is incompatible with the traditional view (e.g. Howard 1971) that the large-scale field patterns arise only from the decay of active region fields.

The evolution of this and other patterns on a longer (cycle) time scale can be followed in Figure 9.3(b), which shows that the large-scale patterns are long-lived and coherent. Among the most distinct patterns are the narrow diagonals at high latitudes, which dominate the left and right columns. In the north the diagonal is black (negative polarity); in the south it is white (positive polarity). The slopes of these diagonals indicate rotation rates which are not only slower than the Carrington rate (defined for latitude 16°), as expected, but vary with the phase of the cycle. For example, in 1967–69 in the northern hemisphere there appear to be two rotation rates in coexistence. The slower rate (indicated by the flatter diagonal) terminates at the end of 1968, but the faster can be followed clearly through 1971. Similar dual rotation-rate patterns can be seen in the southern hemisphere in 1968 and 1970. It would seem that these patterns are controlled by mechanisms which rotate at different rates.

9.4 The polar crown gaps

McIntosh (1981) has emphasized the importance of the behaviour of the chain of filaments which encircle the polar regions in each hemisphere, known as the *polar crown*. Prior to sunspot maximum, each crown is interrupted by a gap, the *polar crown gap*, through which the polar coronal hole extends to lower latitudes.

A study of the complete global maps from which the stackplots of Figure 9.3(b) are constructed reveals that the diagonal patterns are, in fact, traced by the polar crown gaps. Just prior to sunspot maximum, however, the polar crown gaps close and the polar crowns move rapidly polewards, although not simultaneously. This can be seen in Figure 9.3(b), where the diagonal patterns formed by the dark features in the north and the light features in the south terminate abruptly in mid-1968, setting the stage for the polar field reversals.

Interestingly, the slopes of these diagonals decrease, reaching an asymptote representing a slowest rotation rate just before the pattern terminates. This represents an evolutionary discontinuity just prior to sunspot maximum which is clearly a critical point in the sunspot cycle, and it appears that this discontinuity involves a change in the distribution of coronal holes. Other discontinuities can be seen, particularly in the south, where they mark the end of the maximum phase in 1971. McIntosh (1992) has noted similar discontinuities during Cycle 21 and at the beginning of Cycle 22, each of which is marked by an abrupt change in the rotation rate.

9.5 The beginning of Cycle 22

Unfortunately, the Hα observations made at the start of Cycle 20 cannot be compared directly with synoptic magnetic data, but if such behaviour is typical of the growth of the large-scale fields, it should be possible to detect similar patterns at the beginning of the present cycle, for which magnetic data are available from many sources. Accordingly, the National Solar Observatory (NSO) daily magnetograms and synoptic magnetic charts from both NSO and the Mount Wilson Observatory (MWO), together with the Hα synoptic charts produced by the Space Environment Laboratory of NOAA, have been examined to see if corresponding examples can be found at the beginning of Cycle 22.

The situation was not as simple as that in the southern hemisphere at the beginning of Cycle 20. New-cycle fields emerged while old-cycle fields were still present and, in some cases, the distinction between old- and new-cycle regions was not entirely clear. It was nevertheless possible to follow the evolution of two trans-equatorial, large-scale regions and, by using a simulation technique based on the flux transport equation, to assess the contributions of the decaying active regions to their development.

These studies have been described by Wilson and McIntosh (1991) and by Wilson (1992). Because they are of importance to the development of a new approach to the evolution of the large-scale fields, they are discussed in some detail here. The more casual reader may, however, wish to skip directly to the summary in § 9.6.

The development of a large-scale region of trans-equatorial field can be seen in the partial NSO synoptic magnetic charts for the longitude range CL 150° to 270° and for selected Carrington rotations between 1774 and 1781, which are shown in Figure 9.4. The active regions appearing in the upper panels are identified by serial numbers assigned by the Space Environment Services Center of the NOAA Space Environment Laboratory; the regions indicated by letters were not assigned numbers in the NOAA system but are further discussed below.

The development of this large-scale pattern can also be seen in Figure 9.5, where synoptic contour maps for the slightly larger longitude range CL 130°–300° have been constructed from the data provided by the MWO for CRs 1774–79, and in Figures 9.6 and 9.7 for CRs 1780–86. Also shown in Figures 9.6 and 9.7 are contour maps derived from simulations of the fields which are described below. Because these maps are of lower resolution than the NSO maps (4° in longitude and latitude), it is

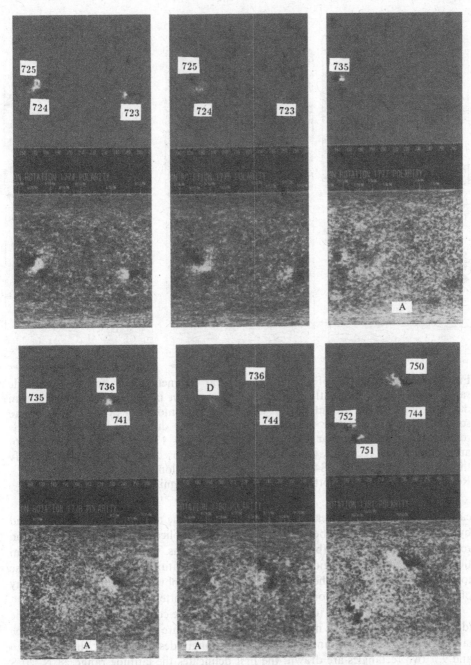

Fig. 9.4 A montage of partial synoptic charts of magnetograms from the Kitt Peak station of the National Solar Observatory (NSO) for CRs 1774–1781 and the longitude range from CL 150° to 270°. The upper panels represent the relative intensity of the magnetic fields and the lower panels the net polarity of the field per pixel. Black is negative polarity, white is positive polarity. The numbers near active regions are the last three digits of the serial numbers assigned by NOAA. The letter 'D' labels a bipolar region that was unspotted and, therefore, did not receive a serial number. The letter 'A' identifies the region A which is discussed in the text.

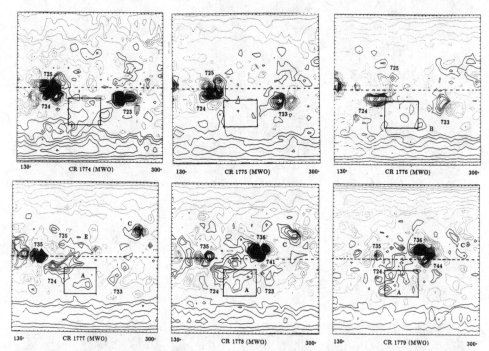

Fig. 9.5 Contour charts of the observed solar magnetic fields for the longitude range 136° − 296°, a slightly larger area than shown in Figure 9.4, derived from data provided by the Mount Wilson Observatory. Contours of positive polarity are solid, contours of negative polarity are dashed. The difference between the fields represented by each contour level is 0.25 G. CRs 1774–1779 (May–September 1986) are included. The rectangles drawn in these and in subsequent contour plots identify the latitude range S12°–S40° and the Carrington longitudes (CL) 186° − 222°. Region C is another unspotted bipolar region with significant flux.

easier to locate the large-scale field regions. Again, the active regions are identified by their NOAA serial numbers, and these are also applied in subsequent Carrington rotations to the enhanced (but unspotted) field regions, which appear to arise from the decay of numbered regions. Distinct magnetic features, which did not give rise to sunspots on the visible hemisphere and were not obviously related to the decay of the identified active regions, are indicated by the letters A to L.

For reference, rectangles (or boxes) are drawn on the contour plots to facilitate the identification of particular magnetic features. In Figure 9.5, the box delineates the latitude range 12°S–41°S and the Carrington longitudes 190°–230°. In Figures 9.6 and 9.7, two rectangles are drawn; the first delineates the latitude range 28°N–20°S and Carrington longitudes 190°–250°, and the second the latitude range 20°N–44°S and longitudes 250°–286°.

The contour maps of Figure 9.5 show that, in CR 1775, the longitude range indicated by the box was essentially clear of significant large-scale field structures, but, between CRs 1776 and 1779, a large-scale pattern of positive polarity field

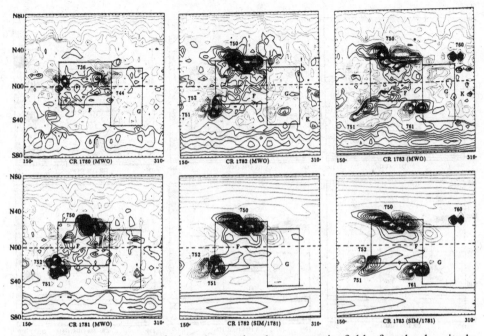

Fig. 9.6 Contour charts of the observed solar magnetic fields for the longitude range 150° − 310° derived from data provided by the Mount Wilson Observatory and labeled (MWO). Contours of positive polarity are solid, contours of negative polarity are dashed. The difference between the fields represented by each contour level is 0.25 G. CRs 1780–1783 (16 September 1986–3 January 1987) are included. Contour charts for CRs 1782 and 1783, derived from the simulated evolution of the large-scale solar magnetic fields based on observed magnetic fields for CR 1781 using the flux transport equation, are shown immediately below the observed charts for these rotations and are labelled (SIM/81). For reference, the two rectangles drawn on the contour plots identify: first, the latitude range N28°−S20° and the CLs 190° − 250°, and second, the latitude range N20°−S44° and the longitudes 250° − 286°.

developed within it. By CR 1780, this pattern, identified by the letter 'F' in Figure 9.6, extended from latitude N30° to the southern polar crown near S50°, the boundary to the positive polarity region surrounding the south pole. This region can also be seen clearly in the Hα synoptic charts for the same period. A region marked 'A', which first appeared within the box in CR 1777 (see Figure 9.5) and can be followed through CR 1780, is discussed below.

In CR 1781, the first major active regions of the new cycle in this longitude range, Region 750 in the northern hemisphere and Regions 751 and 752 in the south, emerged on diagonally opposite sides of region F. This region continued to strengthen through CRs 1782 and 1783, forming a 'flux bridge' connecting the decaying follower flux of Region 750 to the, now combined, leader flux of regions 751 and 752 through CR 1784, after which it began to lose its coherence.

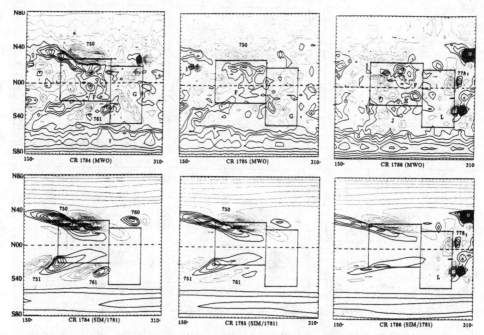

Fig. 9.7 As for Figure 9.6 for CRs 1784–1786 (4 January–26 March 1987).

Region F can also be seen in the longitude range CL 180° − 240° in the Hα maps for CRs 1781 and 1782, which are shown in Figure 9.8, and the development of a coronal hole can be seen in the middle of the region in CR 1782. The large-scale patterns shown by the Hα maps for the period correspond closely with those shown by the magnetic contour maps. This lends credence to the inferences regarding the evolution of the development of the magnetic fields at the beginning of Cycle 20, for which these maps are the prime source.

Unlike the beginning of Cycle 20, when only a single small active region emerged at the time of the development of the large-scale field patterns, several active regions, such as 723, 724, 725, 736, and 741, emerged during the development of this large-scale pattern. Whereas at the start of Cycle 20 the active region could be ruled out on qualitative grounds as the sole source of the flux which made up the large-scale patterns, the question is not so clear-cut at the beginning of Cycle 22.

Stackplots derived from these maps indicate that these active regions did not create an entirely new pattern. Figure 9.9 shows that region F formed part of a diagonal pattern which had existed continuously since CR 1750 (July 1984). This feature is indicated by arrows at the top and bottom of the stackplot. In 1986 it appeared to weaken (CR 1773) but was revived by the emergence of flux between CRs 1774 and 1779 and became the site of the first major region of the new cycle in CR 1781.

In order to assess the contribution of the active regions to the new patterns, the observed contours may be compared with those derived from simulated field

(a)

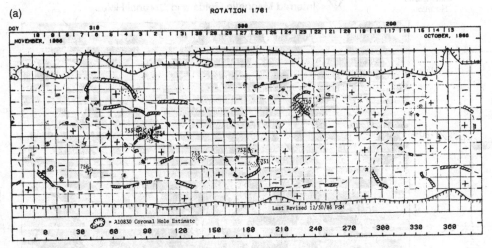

Heliographic Longitude

(b)

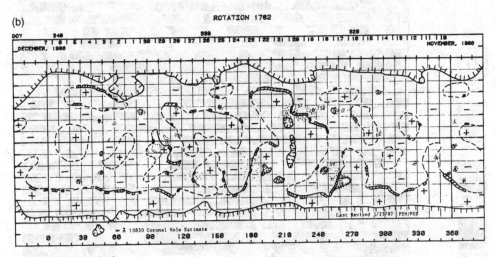

Heliographic Longitude

Fig. 9.8 Hα synoptic charts of magnetic fields for CRs 1781 (13 October to 10 November 1986) (upper) and 1782 (10 November to 7 December 1986) (lower) are shown. The upper numbers show the days of year, with the dates of observations below. Note the development of a coronal hole in CR 1872 in the middle of the region marked 'F' in Figure 9.6.

distributions. Following DeVore and Sheeley (1987), the flux transport equation, Equation (5.1), is solved numerically, in the cartesian approximation, taking as initial conditions the observed MWO magnetic field distributions for selected rotations. In

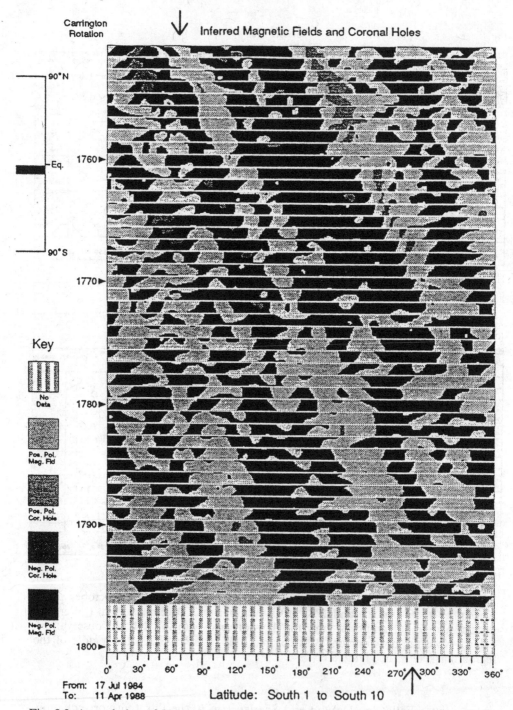

Fig. 9.9 A stackplot of Hα synoptic charts for the latitude zone S1°-S10° for CRs 1750–1800 is shown. The arrows indicate the long-term pattern within which Region F evolved.

these calculations the surface component of the large-scale velocity field is

$$\mathbf{v_s} = v_{dr}\mathbf{i} + v_{mf}\mathbf{j},$$ (9.1)

where v_{dr} describes the component due to differential rotation relative to the rotation rate for latitude $\lambda = 16°$, v_{mf} represents the component due to meridional flow, and **i** and **j** are unit vectors in the azimuthal and meridional directions respectively.

Using the expression derived by Snodgrass (1983), i.e.

$$\omega(\lambda) = 2.902 - 0.464 \sin^2 \lambda - 0.328 \sin^4 \lambda$$ (9.2)

(in μrad s^{-1}), v_{dr} is given in m s^{-1} by

$$v_{dr}(\lambda) = R \cos \lambda \{\omega(\lambda) - \omega(16°)\},$$ (9.3)

R being the solar radius.

For v_{mf}, the reference model of DeVore and Sheeley (1987) was used, i.e.

$$v_{mf}(\lambda) = 10 \left\{ \frac{100^{50}}{99^{49.5}} \cos^{99} \lambda \sin \lambda \right\}^{0.04}.$$ (9.4)

In this model the velocity is zero at the equator, reaches a poleward maximum of 10 m s^{-1} at latitude 5.7° in either hemisphere, and has an angular width of 40°. Following DeVore and Sheeley, the kinematic diffusivity κ was taken as 600 km^2 s^{-1} for the initial calculations.

In the first series of simulations, the MWO data from CR 1774 were used to provide the initial configuration. Each rotation was constructed using 60 time intervals of ∼ 11 hours each, and a grid resolution of 4° × 4°. The grid values so obtained were then used as the initial conditions for the next rotation. Newly emerging active regions were incorporated into the simulations by a technique described by Wilson and McIntosh (1991). The contours derived from these simulated grid values for CRs 1775 to 1780, based on the observed values for 1774, are shown in Figure 9.10.

In CR 1781, the first major active regions of the new cycle emerged, and a new series of simulations of the fields for CRs 1782–86, based on the observed field distribution for CR 1781, was begun. Contour charts, derived from these simulations, are compared with those based on the observed distributions in Figures 9.6 and 9.7. Fields related to new active regions 760 and 761 were included, as in the first series of simulations.

At first glance, the observed contours for CRs 1779 and 1780 and the corresponding simulated contours based on CR 1774 appear to have little in common (apart from the newly emerging regions), and the same is true for those of CR 1786. A detailed comparison, made by Wilson and McIntosh (1991) and Wilson (1992), revealed several significant differences.

(i) The fine structure, which is evident in all the MWO contours, particularly at latitudes poleward of the polar crown, is eliminated from the simulated maps after one or two rotations. This can be seen by comparing the contours of the observed polar fields for CR 1781 with the simulated contours for CRs 1782 and 1783 in Figure 9.6.

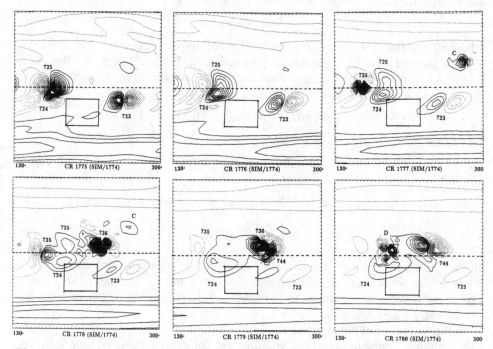

Fig. 9.10 Contour charts are derived from the simulated evolution of the large-scale solar magnetic fields, based on observed magnetic fields for CR 1774, using the flux transport equation. The emergence of new active regions is numerically introduced into the maps as described in the text. The contour spacings are as for Figure 9.5.

(ii) The upper half of the box in the observed contours for CR 1780 contains the well developed large-scale field, which is labelled F in Figure 9.6 and was the site of the major new-cycle regions which emerged in CR 1781. Although it is likely that it derived, at least in part, from the decay of the old-cycle regions which had emerged in this location since CR 1774, the simulated contours in Figure 9.10 show that the decaying remnants of these fields intrude only into the top left corner of the box.

(iii) The emergence and maintenance of one region, marked as A in CRs 1777–80, and another as G in CRs 1782–86, cannot be seen in the simulated contours; a careful study of the daily magnetograms shows that they cannot be related to the decay of any observed active regions.

9.6 Discussion

The rapid elimination of the fine structure in the simulated contours is not surprising, because the diffusion term is a major component of the flux transport equation. Since the equation contains no source terms, there is no way in which the fine structure can be introduced into the simulated patterns. This comment, however, serves to emphasize that small-scale magnetic regions, in the form of ephemeral and other small active regions, are continually emerging and are widely distributed, both in longitude and in latitude (Harvey and Martin 1973). It is very likely that this emerging small-scale flux is responsible for the maintenance of the fine structure of the magnetic patterns. However, an even more important consequence of the *non-random* emergence of small bipolar flux regions is the creation of large regions of apparently unipolar flux, which is discussed below.

Independently, Stenflo (1992) has argued that the turnover time of the large-scale field (i.e. the time in which the individual flux elements decay and are replaced by new elements within the pattern) is of the order of a week. This is two orders of magnitude shorter than the time implied by the Babcock–Leighton model for the evolution of the large-scale fields by active region decay. Noting that the *total* flux emerging in the smaller elements exceeds that emerging as large active regions by several orders of magnitude, Stenflo has proposed that the high- or mid-latitude large-scale fields arise primarily as a result of organized small-scale flux emergence rather than active region decay.

It would appear that regions A and G provide specific examples of this phenomenon. A mechanism which can account for this, without violating the zero divergence of magnetic fields or postulating the existence of magnetic monopoles, is described in §9.9.

9.7 The beginning of a new cycle

These observations also show that the transition from Cycle 21 to 22, in the longitude range CR 150°–300°, is closely related to the development of the cell-like pattern of large-scale field, dominated by a large region of positive flux.

Both old-cycle and new-cycle active regions are associated with the neutral lines of the cell and, in some cases, the distinction between old- and new-cycle regions is uncertain. Because of its latitude and orientation, AR 723 is clearly old-cycle, southern hemisphere, but AR 736, which emerged at the same longitude in CR 1778 at low latitude in the northern hemisphere, has the same polarity orientation and therefore is either an unusual low-latitude, new-cycle, northern hemisphere region or an old-cycle region from the southern hemisphere which has drifted across the equator.

K. Harvey (1992) has discussed this problem and argued that, prior to the rise in the number density of spots near the equator as the new cycle moves into the zone, all equatorial spots should be regarded as belonging to the old cycle, irrespective of orientation. AR 736 should therefore be regarded as an old-cycle region, but it is of interest that AR 750, which is obviously a new-cycle region, appears at the northern extension of the *same* neutral line in CR 1781.

Again, AR 761, which emerged in the northern hemisphere during CR 1783, is clearly a new-cycle region, but, in CR 1786 (see Figure 9.7), AR 778 emerged at low latitude in the northern hemisphere, and a region marked 'L' appeared in the south. According to its orientation AR 778 should be classified as an old-cycle northern region, but it straddles the same neutral line as Region L, which is clearly a southern region of the new cycle. While it is possible that the emergence of these regions along the same meridian is coincidental, there is a tendency for active regions to emerge across the neutral lines of existing large-scale fields, and it would seem that the formation of a trans-equatorial cell such as this plays a significant role in the transition from Cycle 21 to Cycle 22. Similar patterns can be found at the beginnings of Cycles 20 and 21.

9.8 The origin of the large-scale fields

The observations discussed above offer a very different view of the development of the large-scale field patterns from that given by the kinematic model. In place of a simple 'cause and effect' relationship between the active regions and the large-scale fields, they appear to have a common origin which may be related to the deep-seated velocity fields of the convection zone. The long term persistence of the patterns in Figures 9.3(b) and 9.9 lends support to this interpretation. However, if large-scale field patterns may appear without the prior emergence of active regions, some account of this process must be offered.

Although large-scale unipolar regions may include many small bipoles, it is generally assumed that they also contain field elements of a single polarity which are the foot-points of large-scale magnetic field structures projecting through the photosphere, linking this region to another unipolar region of opposite polarity. As discussed in §9.2, the magnetograph registers a (positive) unipolar signal because it records an average over both the small

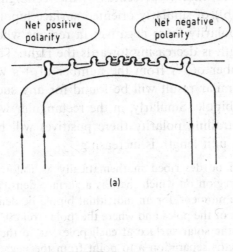

(a)

(b)

Fig. 9.11 (a) The sketch illustrates (in vertical section) how an inhomogeneous distribution of aligned bipoles may arise in relation to a subsurface Ω loop. (b) The diagram shows (in highly idealized surface projection of the structure shown in (a)) an inhomogeneous surface distribution of aligned bipoles, indicated by (+−) pairs, such that the net flux within the rectangle (window) on the left is positive, while the net flux within that on the right is negative. The net polarity of the whole region, however, is zero.

bipoles and the foot-points within the 'surface slice' through the large-scale field structures. However, Snodgrass and Wilson (1993) have shown that this is not the only possibility, and that a unipolar flux signal may also arise where there is a local gradient in the surface density of unresolved but preferentially oriented magnetic features at the surface. In terms of this result, the observations cited above may be interpreted without violation of the laws of physics.

A simplified picture that illustrates the basic principle is seen in Figure 9.11. Figure 9.11(a) shows a vertical section of an idealized, non-uniform distribution of small Ω-loops which are connected to a sub-surface element of toroidal field, while Figure 9.11(b) shows an idealized horizontal section of the bipoles, represented by (+−), which are the foot-points of the Ω-loops.

A low-resolution observation (represented by the rectangular window on the right of Figure 9.11(b)) of similarly oriented surface bipoles would record a net flux of leading polarity (here negative) in regions where the number of bipoles per unit length is decreasing towards the right. This can be seen by subtracting the number of −'s from the number of +'s within the window and noting that a similar result will be found for any such window of size larger than a single bipole. Similarly, in the rectangular window on the left, a net flux with the trailing polarity (here positive) will be measured since here the number per unit length is increasing.

This phenomenon can be described mathematically as follows: consider a two-dimensional (surface) region on which there is a surface density of small magnetic bipoles. Let the 'bipole moment' for an individual bipole be defined as $\mathbf{a}\phi$, where $\mathbf{a}$ is the vector separation of the poles and where the 'pole strength' ϕ is the magnitude of the flux normal to the solar surface at each pole. As in the case of an electric dipole, we take the vector separation $\mathbf{a}$ to point from the negative pole, where the flux passes into the surface, to the positive pole, where it passes out.

Let $n(\mathbf{r})$ be the bipole density, i.e. the number of bipoles per unit area at the point $\mathbf{r}$. Denote by $\delta n(\mathbf{r})$ the bipole density for those bipoles with pole separation between $\mathbf{a}$ and $\mathbf{a} + d\mathbf{a}$, and pole strength between ϕ and $\phi + d\phi$, and let 'δ' have a similar connotation when acting on other quantities. Thus the total bipole moment per unit area, $\mathbf{b}(\mathbf{r})$, is found by integrating $\delta\mathbf{b}(\mathbf{r}) = \mathbf{a}\phi\delta n(\mathbf{r})$ over all pole strengths and (vector) pole separations.

We represent the number densities of positive and negative poles separately as $\delta n_+(\mathbf{r})$ and $\delta n_-(\mathbf{r})$ respectively. Then the contribution to B_r, the net flux per unit area passing normal to the solar surface from the bipoles of pole strength ϕ and separation $\mathbf{a}$, is:

$$\delta B_r(\mathbf{r}) = \phi\delta n_+(\mathbf{r}) - \phi\delta n_-(\mathbf{r}). \tag{9.5}$$

Since each of these bipoles consists of a positive pole and an equally strong negative pole, separated by the vector $\mathbf{a}$,

$$\delta n_+(\mathbf{r} + \mathbf{a}) = \delta n_-(\mathbf{r}). \tag{9.6}$$

Thus to first order in a Taylor series,

$$\delta n_+(\mathbf{r}) - \delta n_-(\mathbf{r}) = -\mathbf{a} \cdot \nabla\delta n_+(\mathbf{r}), \tag{9.7}$$

and since, also to first order, $\delta n_+(\mathbf{r}) = \delta n(\mathbf{r})$, we obtain:

$$\delta B_r(\mathbf{r}) = -\phi\mathbf{a} \cdot \nabla\delta n(\mathbf{r}). \tag{9.8}$$

If neither $\mathbf{a}$ nor ϕ depends on position $\mathbf{r}$, we obtain, on integrating over $\mathbf{a}$ and ϕ:

$$B_r(\mathbf{r}) = -\nabla \cdot \mathbf{b}(\mathbf{r}). \tag{9.9}$$

Thus, where there is a divergence in the bipole moment density arising from a gradient in the number density of bipoles along the preferred direction of dipole orientation, the region will exhibit a net magnetic polarity, negative or positive respectively, when the gradient (or divergence) is positive or negative.

This result corresponds to the relationship between 'magnetic pole density' and the 'magnetization' in the theory of magnetic materials (cf. e.g. Reitz and Milford 1960). It is also analogous to the creation of surface charge on a polarized dielectric.

The resulting apparently unipolar region is called a *virtual unipolar region* to distinguish it from a unipolar region which includes the foot points of field lines extending into the chromosphere and corona. In practice, it is unlikely that a unipolar region observed on the Sun is strictly a virtual unipolar region; there are always likely to be some foot-points of coronal fields present. However, it is equally unlikely that unipolar regions consist only of such foot-points; high-resolution magnetograms always show the 'pepper and salt' pattern indicating the presence of small bipoles, and the above analysis shows that, if these bipoles have an appropriately inhomogeneous distribution, they may contribute to the unipolar signal in a low-resolution magnetograph. It may not be possible to distinguish between these two types of region by magnetograph measurements alone.

Just as for real fields, however, the virtual fields must cancel when summed over the the whole solar surface. This follows immediately by integrating Equation (9.5) over an area sufficient to contain all the bipoles in the system.

We now consider the circumstances under which virtual unipolar regions may appear on the Sun and how they can provide some insight into the phenomena described above.

9.9 Sources of organized bipole distributions

There are, in general, two processes which may give rise to the appearance of small magnetic bipoles at the solar surface. In the first, suggested by Stenflo (1992), small disconnected loops of field (*ring loops*) are shredded from a deep-lying flux rope. These ring loops will rise under buoyancy through the convection zone and are subject to the competing effects of expansion, due to the density and pressure gradients, and contraction under their own magnetic tension (Wilson *et al.* 1990). Although these loops will suffer Coriolis twisting as they rise, those that reach the surface as bipoles may retain some trace of the orientation imposed on them by the deep flux rope from which they originated.

Although the lifetime of each such bipole may be short, Stenflo emphasizes that the number of bipoles which appear at the surface at the small-scale end of the spectrum (seen as the 'salt and pepper' pattern in high-resolution magnetograms) is so large that even a small bias in their orientations should be significant. If, in addition, they exhibit a density gradient in the preferred direction, the above analysis shows that virtual unipolar regions may appear, disappear, and move about as the distribution of these loops changes with time. In this case, the magnetograph will register field patterns in which the unipolar regions appear to have little relationship to each other, although the net flux from all of them over a sufficiently wide region must sum to zero, as noted above.

An alternative process involves the field rising to the surface as an Ω-loop which is not disconnected from the underlying field structure. Although a large active region emerges as an eruption through the surface of a twisted, and therefore concentrated, Ω-loop of flux, there would appear to be no *a priori* reason why some Ω-loops should not be less twisted and more diffuse than the loops that produce the large-scale active regions. These loops would not thrust themselves aggressively through the surface but would escape more slowly, decaying by the *sea-serpent* process (Spruit *et al.* 1987) through the development of Kelvin–Helmholtz instabilities and the interaction with small-scale velocity eddies. This process would produce a distribution of bipoles at the surface which would be organized both in density and average orientation by their relation to the diffuse sub-surface toroidal element.

Evidence for such bipole distributions at high latitudes has been described by Martin and Harvey (1979) and is discussed in Chapter 8. In terms of the EAC, it suggests that these more diffuse toroids occur at high latitudes during the declining phase of the cycle.

9.10 Development of a 'real' unipolar region

Snodgrass and Wilson (1993) also show how a pair of virtual unipolar regions which appear as a result of the second process may evolve into a genuine large-scale field pattern with unipolar regions connected above the photosphere in a manner consistent with observations. For a field pattern such as that seen in Figure 9.11(a), studies of ephemeral regions (Martin and Harvey 1979) indicate that the lifetimes of small bipoles are of the order of days or less. They appear to decay by reconnection with similar adjacent bipoles or with network elements, as illustrated by the dashed lines in Figure 9.11(a) and the corresponding reconnected lines in Figure 9.12(a). These reconnections between adjacent loops leave behind small ring loops which rapidly decay (Spruit *et al.* 1987), and the subsurface toroid thus 'leaks' through the surface.

As more bipoles emerge, the reconnections continue. Many of these will occur between bipoles of random orientation, but eventually, because there

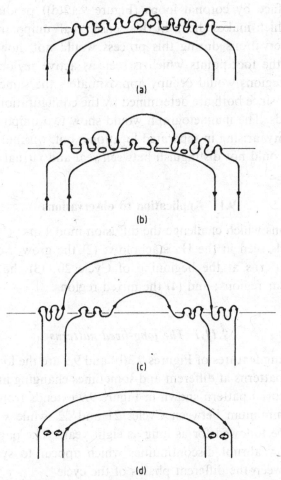

Fig. 9.12 The sketches illustrate (in vertical section) four stages of the evolution of the configuration shown in Figure 9.11(a) to two large-scale unipolar regions of opposite polarities connected by a coronal loop.

is a preferred orientation, the cancellations will result in a chromospheric arcade between the virtual field regions. This is illustrated in Figure 9.12(b). The neutral line of the configuration should be visible as a filament in Hα.

After some time, the emergence of bipoles at the middle of the region declines, but the reconnections continue until the total flux intersecting the surface in this middle region is less than that at either end (see Figure 9.12(c)). Further, there is now a real excess of flux of leader or follower polarity in the now isolated end regions.

Eventually, the excess flux elements of opposite polarities are connected above the surface by coronal loops (Figure 9.12(d)) produced by the reconnections which makes them, in this sense, 'real' unipolar regions. Magnetograph observations during this process would not, however, show the separation of the foot-points which is seen as active regions emerge. The real unipolar regions would occupy approximately the same regions as the virtual regions, since both are determined by the configuration of the original subsurface fields. The magnetograph would show two unipolar regions with opposite polarity, arising in separated locations; but, without concurrent Hα filtergrams, it could not distinguish between real and virtual regions.

9.11 Application to observations

The observations which challenge the diffusion model are: (1) the long-lived patterns or cells seen in the Hα stackplots; (2) the growth of the southern hemisphere patterns at the beginning of Cycle 20; (3) the appearance of isolated unipolar regions; and (4) the mixed regions.

9.11.1 The long-lived patterns

The most striking features of Figures 9.3(b) and 9.9 are the long-lived streaks traversing the patterns at different and sometimes changing inclinations. The principle equatorial pattern shown in Figure 9.9 extends from 1984 to 1988, spanning the minimum between Cycles 21 and 22, while some large-scale patterns may be followed for as long as eight years. The patterns of Figure 9.3(b) also show abrupt discontinuities which appear to synchronize with transitions between the different phases of the cycle.

The inclination of the streaks in the stackplots indicates the rotation rate of whatever agency maintains these patterns over the long term. Individual surface features may have different rotation rates, but, having much shorter lifetimes, they may drift in and out of the long-lived pattern without disturbing its inclination. (See Snodgrass (1992) for a further discussion.) On the assumption that the source of these fields is located at some level below the surface, the inclination of the streak probably represents the rotation rate at that depth and latitude. Changes in the inclination may indicate changes in the rotation rate at that level or a change in the level of the source which continues to replenish the surface fields. Intersecting streaks indicate that some surface features at the same latitude may arise from sources at different depths.

If the large-scale field patterns arose only from passive diffusive decay of active-region fields, one might expect that their rotation rates at given latitudes would be determined only by the mean differential rotation rate for that latitude. The multiple rotation rates exhibited by these long-lived features suggest the presence of large-scale, sub-surface, non-axisymmetric agencies, such as velocity fields or cells, which continuously expel flux towards the surface, perhaps in the form of a loosely wound Ω-loop as described in §9.9, the neutral lines of the field delineating the boundaries of the velocity cells. Further, since the large active regions tend to emerge across the boundaries of the cells, the same velocity cells may also play a role in the emergence of active regions. The longevity of the patterns supports the existence of some large-scale sub-surface agency, and the abrupt discontinuities which occur in the patterns at the transitions between different phases of the cycle indicate that this agency plays an important role in the cyclic process.

9.11.2 The southern hemisphere fields of Cycle 20

Of the several instances in which the appearance of large-scale patterns of positive and negative field cannot be related to the decay of active regions, perhaps the most striking example occurred at the beginning of Cycle 20, which was discussed in §9.3 above. At that time the southern hemisphere remained essentially free of large-scale field patterns and of major active regions until Carrington rotation 1512, when the first persistent large-scale patterns appeared at the same time as a very small sunspot group. It was not, however, until CR 1515 that the first major active region of the cycle for that hemisphere emerged, straddling a neutral line of the pre-existing pattern.

We hypothesize that many of the subsurface Ω-loops which formed in the southern hemisphere at the beginning of Cycle 20 were diffuse, rather than compact, and that these gave rise first to virtual unipolar field regions, which then evolved into conventional unipolar field regions connected by coronal field loops. It is thus possible to understand how these large-scale flux patterns emerged *before* the emergence of the first major active regions. Further, if whatever process generated the diffuse loops also generated the more compact active-region loops, it is not surprising that the first major active region of that cycle and hemisphere emerged across a neutral line of the already established large-scale field (McIntosh and Wilson 1985). This is an example of the 'nesting' of regions (§3.11) which may include those giving rise directly to the large-scale fields.

Examination of synoptic maps of the magnetic field near the beginning of more recent cycles (see §9.5) shows that, during the early stages of a cycle, it is not uncommon for the background field neutral lines to identify the longitudes where the new-cycle active regions of the same polarity orientations tend to develop (Wilson 1992).

9.11.3 Isolated unipolar regions

While the flux associated with the virtual large-scale fields must balance when integrated over the entire solar surface, as does the flux from the real fields passing through the surface, it may, for a number of reasons, not be possible to identify the balancing regions in the magnetograms.

As discussed in §9.2, the magnetograph saturates at high field strengths and has a weak but variable bias which interferes with weak-field measurements. This bias arises from sources such as polarization produced in the optics and imperfect cycling of the optically active crystal. In the virtual unipolar field situation the bipole density gradient may not be the same on both sides of the 'neutral' line, and the virtual unipolar regions may be separated by a region in which the dipole density is roughly constant. The appearance of the isolated unipolar regions identified as 'A' and 'G' at the beginning of Cycle 22 (see §9.5) may thus be due to the magnetograph's response to some variability in the bipole density gradient on either side of the neutral line, and the absence of obvious balancing regions of unipolar flux in these magnetograms need not be unduly puzzling.

9.11.4 Mixed regions

In the other cases reported (e.g. Region F and some of the polar field patterns (Murray and Wilson 1992) to be discussed in Chapter 10), the fields from decaying active regions cannot account completely for the large-scale field patterns but have obviously made a significant contribution. In these cases it would seem that the large-scale unipolar field signal registered by a magnetograph may arise from two sources: (i) the isolated foot-points of coronal loop structures and (ii) a non-uniform distribution of unresolved surface bipoles having a preferred alignment. These sources have been characterized as the 'real' field component and the 'virtual' field component.

Any magnetogram may include contributions from each of these components, and there is no reason why the two processes should be regarded as entirely distinct. They simply lie at opposite ends of the size spectrum of emerging bipolar regions, and it would seem likely that in the evolution of some large-scale field regions contributions are made from bipolar regions of all sizes.

9.12 Conclusion

The development of the large-scale trans-equatorial field patterns at the beginnings of all three recent cycles is very different from the standard picture of the growth of the large-scale fields. These observations show that, rather than arising solely from the diffusive decay of the AR fields, both the large-scale fields and the ARs develop as part of 'active cells' which may have lifetimes of several years.

While the decay of these active-region fields undoubtedly contributes to the large-scale field patterns of the cell, there is evidence from Cycles 20 and 22 that at least part of the large-scale flux emerges independently of the decay of the active regions, the coherence of the cell structure suggesting a common origin for both categories of field.

The mechanism described in §9.9 indicates that both loosely and tightly wound flux ropes may be present in the structure of the cell described at the start of Cycle 22, whereas, at the beginning of Cycle 20, only the more loosely wound ropes were present initially.

It is also found that active regions of both old and new cycles emerge across the neutral lines of the cell, in one case across the same trans-equatorial neutral line and, in another, within the same hemisphere, a result which emphasizes the importance of the neutral-line structure and suggests a link with deep-seated dynamical configurations. Further, in both Cycles 20 and 22, the transition from old to new cycle was marked by the appearance of a northern-hemisphere region which was of abnormally low latitude for a new-cycle region but of the wrong polarity orientation for the old cycle.

If these were, in fact, old-cycle regions, carried across the equator by subsurface velocity flows, these flows may be related to the development of the cell and to the establishment of the new cycle.

The significant retardation of the rotation rate of the patterns shown in Figure 9.3(b), just prior to the closure of the polar crown gap and the abrupt change of the pattern structure, also points to a change in the subsurface velocity fields. This may be related to the reversal of the polar fields and is further discussed in the next chapter.

References

DeVore, C. R., Sheeley, N. R., and Boris, J. P.: 1984, *Solar Phys.*, **92**, 1–14.

DeVore, C. R., and Sheeley, N. R.: 1987, *Solar Phys.*, **108**, 47–59.

Harvey, K. L.: 1992, in *Proceedings of the National Solar Observatory/Sacramento Peak 12th Summer Workshop*, ed. K. L. Harvey, San Francisco, California, 335.

Harvey, K. L., and Martin, S. F.: 1973, *Solar Phys.* **32**, 389.

Howard, R. F.: 1971, *Pub. Astron. Soc. Pacific*, **83**, 550.

Howard, R. F.: 1984, *Solar Phys.*, **39**, 275.

Howard, R. F.: 1984, *Ann. Rev. Astron. Astrophys.*, **22**, 131.

Howard, R. F., Bumba, V., and Smith, S. F.: 1967, *Atlas of Solar Magnetic Fields*, Publication 26, Carnegie Institution of Washington.

Howard, R. F., and LaBonte, B. J.: 1981, *Solar Phys.*, **74**, 131.

Howard, R. F. and Stenflo, J. O.: 1972, *Solar Phys.*, **22**, 402.

McIntosh, P. S.: 1972a, *Solar Activity Observations and Predictions*, Vol. 30 of *Progress in Astronautics and Aeronautics*, ed. M. Summerfeld, Academic Press, New York, 65–92.

McIntosh, P. S.: 1972b, *Rev. Geophys. and Space Phys.*, **10**, 837–46.

McIntosh, P. S.: 1979, *Annotated Atlas of Hα Synoptic Charts*, Upper Atmosphere Geophysics Report 70, NOAA World Data Center A, Boulder, Colorado.

McIntosh, P. S.: 1981, *The Physics of Sunspots*, eds. L. E. Cram and J. H. Thomas, Sacramento Peak Observatory, 7–54.

McIntosh, P. S.: 1992, in *Proceedings of the National Solar Observatory/Sacramen-to Peak 12th Summer Workshop*, ed. K. L. Harvey, San Francisco, California, 14.

McIntosh, P. S. and Wilson, P. R.: 1985, *Solar Phys.* **97**, 59–79.

McIntosh, P. S., Willock, E. C., and Thompson, R. J.: 1991, *Atlas of Stackplots derived from Hα Synoptic Charts for 1966–1987*, Upper Atmosphere Geophysics Report 101, NOAA National Geophysical Data Center, Boulder, Colorado.

Martin, S. F., and Harvey, K. L.: 1979, *Solar Phys.*, **64**, 93.

Murray, N., and Wilson, P. R.: 1992, *Solar Phys.*, **142**, 221.

Reitz, J. R, and Milford, F. J.: 1960, *Foundations of Electromagnetic Theory*, Addison-Wesley, Reading, Mass.

Snodgrass, H. B.: 1983, *Astrophys. J.*, **270**, 288–99.

Snodgrass, H. B.: 1992, in *Proceedings of the National Solar Observatory/Sacra-mento Peak 12th Summer Workshop*, ed. K. L. Harvey, San Francisco, California, 71.

Snodgrass, H. B., and Wilson, P. R.: 1993, *Solar Phys.*, (in press).

Spruit, H. C., Title, A. M., and van Ballegooijen, A. A.: 1987, *Solar Phys.*, **110**, 115.

Stenflo J. O.: 1992, in *Proceedings of the National Solar Observatory/Sacramento Peak 12th Summer Workshop*, ed. K. L. Harvey, San Francisco, California, 83.

Wilson, P. R., McIntosh, P. S. and Snodgrass, H. B.: 1990, *Solar Phys.*, **127**, 1–9.

Wilson, P. R., and McIntosh, P. S.: 1991, *Solar Phys.*, **136**, 221–237.

Wilson, P. R.: 1992, *Solar Phys.*, **138**, 11–21.

10

The reversal of the polar magnetic fields

10.1 The polar fields

The first observations of the Sun's weak polar magnetic fields were obtained in 1915 by Hale at Mount Wilson but, at that time, little attention was paid to their polarities in relation to those of sunspots. In 1957, however, Horace Babcock noted that, at the beginning of Cycle 19, the north polar fields were positive, as were the leader spots of the new cycle in the northern hemisphere. He further noticed that, as the cycle proceeded, the polar fields weakened and, in 1959, the mean magnetic polarity of the north polar field reversed, so that, for a period, the polarity of both polar fields was negative. Eighteen months later, the polarity of the south polar field also changed, so that, at the start of the next cycle (Cycle 20), the polar polarities in each hemisphere again corresponded to those of the leader spots. Similar reversals also occurred shortly after maximum during Cycles 20 and 21. On the basis of these three occurrences, it is now widely assumed that the global polar field of the Sun reverses with a period comparable to that of the sunspot magnetic cycle, but with a phase difference of $\sim 90°$.

The polar fields are weak; even at sunspot minimum they are only a few gauss, and their reversals are not well defined. At times during the reversals, the polar regions are covered with weak patches of field of either polarity and the net polarity of the region is uncertain. The completion of the reversal process, however, seems to 'set the stage' for the new cycle. Shortly after reversal, the large-scale field configuration changes qualitatively to a simpler form, the 10-cm radio flux drops abruptly (McIntosh 1992), and the first small bipolar regions showing polarity patterns consistent with those expected for sunspot pairs of the following cycle and with the (now reversed) polar fields appear at high latitudes (Harvey 1992). The reversal of the polar fields therefore appears to mark a crucial phase of the magnetic activity cycle and must be included in any model which purports to explain cyclic activity.

Clearly, the polar field reversals play an essential role in relaxation models, which require that a global polar poloidal field be established prior to the beginning of each sunspot cycle, with subsurface field lines linking the northern and southern polar fields. In Chapter 6, three heuristic accounts of the polar reversals were discussed: Babcock's (1961) original version, Giovanelli's (1985) modification of that version, and the flux transport model of Leighton (1964), Sheeley (see DeVore and Sheeley 1987), and others.

Each of these models assumed specifically that the poleward drift of the large-scale surface fields is responsible for the reversals and that these fields arise only from the decay of the active region fields. If, however, large-scale flux patterns can emerge at lower latitudes independently of active regions, or jointly as part of long-lived velocity cells, as shown in Chapter 9, then such models are suspect. Indeed, several writers have begun to question the validity of these surface-flux-transport models.

Stenflo (1992) has shown that the flux-transport models are unable to reproduce the rotational properties of the high-latitude fields and argues that the high-latitude fluxes must be continually replenished by sources within the solar interior. He points out that the total flux emerging in the form of small bipoles is several orders of magnitude greater than that emerging in the large active regions, and that only a minor organization or orientation of these fields (such as is discussed in §9.10) would be sufficient to generate the large-scale global field pattern at high latitudes by local flux emergence. Legrand and Simon (1991) and Simon and Legrand (1992) go further and argue for the existence of a two-component cycle in which the dipole (polar) field is generated by a separate mechanism from that which generates the sunspot cycle, although they are linked. They assert that the dipole field is not a surface phenomenon but is deep-seated within the solar interior and linked to the following sunspot cycle with a five to six year delay. In this they appear to endorse the concept of the extended activity cycle (see Chapter 8).

In this chapter the most recent data are examined and these two very different approaches to the polar field reversals are compared.

10.2 The polar crown and the reversals

In Chapter 8 it was shown that the poleward-propagating component of the extended cycle can be identified in data derived from several different phenomena. In synoptic charts constructed from the torsional shear

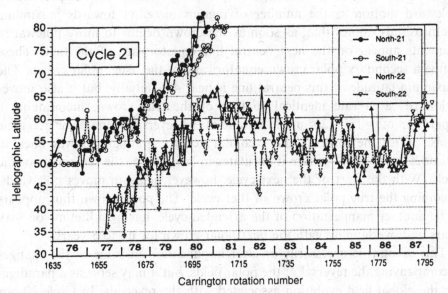

Fig. 10.1 The poleward progress of the primary polar crowns during Cycle 21. Note that the poleward drift of the secondary crown is arrested at the time of the polar field reversals, after which these crowns tend equatorwards in parallel with the phenomena of the EAC.

data (Figure 8.2) and from the green-line emission maxima (Figure 8.3) a poleward-moving branch can be identified in the southern hemisphere, which began at mid-latitudes in 1976–77 and reached the pole in 1980–81, coincident with the polar field reversal. These poleward branches appear to follow the same path as the neutral line in Stenflo's modally clean magnetic butterfly diagram (Figure 8.6).

Another feature which migrates polewards during the rising phase of the cycle and appears to play a part in the reversal process is the polar crown neutral line. One of the longest neutral lines on the solar surface connects the polar crowns of both hemispheres with features near the equator, and this line has been likened to the webbing on a baseball. The polar crowns make up those parts of this line which encircle the poles at the high latitudes. McIntosh (1992) has plotted observations of the latitude of the highest point of this line against Carrington rotation during Cycle 21 (Figure 10.1). Here a uniform poleward migration began in 1976 (CR 1645) near latitude 50° and ended in 1981 at the poles, coincident with the polar field reversals.

McIntosh argues that this steady progress must reflect some fundamental subsurface process rather than the surface diffusion of decaying active-region fields, since the latter should produce a slow poleward drift near minimum, when the number of active regions present is small, and an accelerated poleward motion as the number of regions increases towards maximum. McIntosh also notes that, as soon as the crown begins to move polewards, filaments appear on the next neutral line equatorward of the first. These define a secondary polar crown equatorward of the 'true' polar crown. The maximum latitude of this neutral line is near 40° initially, but it also moves polewards at a rate identical to that of the true crown, maintaining a separation of $\sim 15° - 20°$ until the time of the reversal of the polar magnetic fields. After the reversal, the secondary crown gradually moves back to latitudes $\sim 50° - 55°$, where it oscillates in latitude for the remainder of the cycle. With the start of the next cycle, however, it again moves polewards, becoming the true polar crown of that cycle. The polar crown thus appears to be another manifestation of the extended cycle, having a lifetime of ~ 19 years and overlapping with the next polar crown for up to 8 years.

This account is necessarily an oversimplification of some of the changes accompanying the reversal of the polar fields, but it may serve as a paradigm for the global field evolution associated with the reversals. In Cycle 20, for example, a three-fold reversal occurred in the northern hemisphere as the secondary polar crown followed the first into the polar region, rather than falling back. However, the new polar field (of leader polarity) was weak, and a further reversal occurred, so that, by the time of the next minimum, the reversed (follower) polarity was established. It is not clear whether this three-fold reversal was a random fluctuation of the weak polar fields or a more systematic occurrence. There is some evidence for the occurrence of triple reversals in earlier cycles, and Benevolenskaya and Makarov (1992) claim that triple reversals have occurred during all the even cycles since Cycle 12. They attribute these triple reversals to the superposition of a high-frequency variation on the basic 22-year cycle of the large-scale magnetic field, but they do not suggest a physical mechanism. In view of doubts regarding the reliability of these early records, the resolution of this question must await results from the current and later cycles.

The polarity of the large-scale fields equatorward of the 'true' polar crown is that of the 'follower' flux of the new-cycle active regions, whereas the polarity below the secondary crown is of the 'leader' flux. In order to see whether this configuration arises as the result of the diffusion and random-walk transport of decaying active-region fields, or whether these regions are developed, in part, by non-random local flux emergence, Murray and Wilson (1992) have examined the changes in the polar fields between CRs 1815 and 1834 (July 1988 to October 1990), using magnetic data from the Mount Wilson Observatory (MWO) and the National Solar Observatory

(NSO), and again compared the observed evolution of these fields with that obtained from simulations.

10.3 The high-latitude fields

Figure 10.2 shows the MWO contour maps for rotations 1815 and 1826. In these maps, which use both grey-scale shading and contours, black and broken-line contours represent negative field, white and full lines positive field. While positive flux regions extend up to latitude $\sim N65°$ in CR 1815, both the contours and the grey-scale indicate that, above this latitude, a unipolar region of negative flux occupies the polar region; i.e. this pole has not reversed. By CR 1826, however, the polar region is essentially neutral, and, since the field below latitude $\sim 75°$ is positive, a reversal appears to have taken place. This may not be strictly correct, because the MWO data extend only to latitudes $\pm 75°$, but it is clear that, at least in the range N55°–N75°, a significant polarity change has occurred. The south polar region is, by contrast, weakly positive in CR 1815 and in CR 1826. McIntosh (private communication) has studied the corresponding Hα synoptic maps and, with the benefit of hindsight, reports that the reversal of the north polar field occurred in December 1990 (CR 1837), while the south polar field did not reverse until May 1992 (CR 1854).

In order to follow the change in the north polar field in more detail, the 'stack-plot' technique is used. Here, the range of latitudes from N50° to N75° is extracted from the synoptic map for each rotation and stacked vertically, as shown in Figure 10.3(a). In this way, the positive and negative components of individual large-scale field elements can be followed sequentially through the plot, the leftward drift being due to differential rotation, for which the zero point in Carrington plots corresponds to the mean rotation rate for latitude 16°.

It is clear that, by CR 1826, the unipolar negative field has disappeared from the essentially neutral polar region, whereas the dominant polarity between latitudes N50° − 70° is positive and remains so through CR 1834. Although this period does not see the completion of the reversal process and the establishment of a uni-polar positive field region at the pole, it does cover a significant phase of the reversal.

A similar stackplot was produced for the south polar region and is shown in Figure 10.4(a). Although the polar field is neutral or weakly positive at the beginning and end of this sequence, the positive field weakens after CR 1817 and is replaced by a region of negative field until CR 1824. The negative field then decays and gives way to a positive field.

Again we pose the question: do these field changes arise solely from the poleward drift of fields from the active-region belt or do they emerge, at least in part, *in situ*? At this phase of the cycle, many active regions emerge at lower latitudes, and it is important to be able to determine their effect on the observed changes in the polar fields.

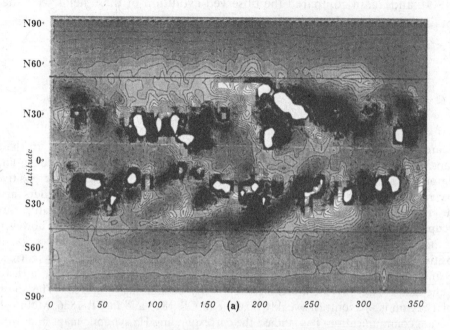

(a)

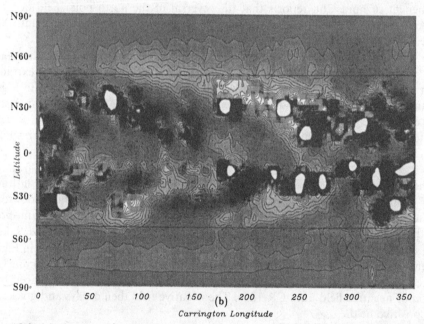

(b)

Carrington Longitude

Fig. 10.2 (a) A synoptic contour chart of the observed solar magnetic fields for Carrington rotation 1815, derived from data provided by the Mount Wilson Observatory. Contours of positive polarity are solid and the region is shaded light, contours of negative polarity are dashed and the regions are dark shaded. The horizontal lines indicate latitude ±50°. (b) A similar map for CR 1826.

Carrington Longitude 0–360   Carrington Longitude 0–360

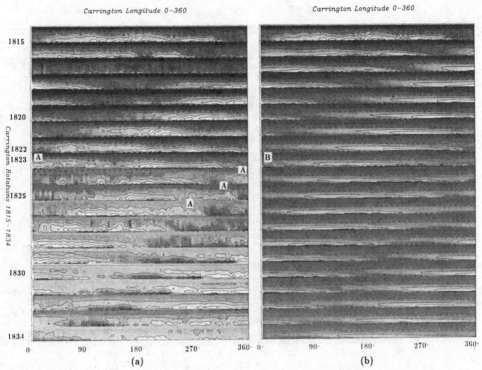

Fig. 10.3 (a) A stackplot of the north polar regions above latitude N50°, constructed from the Mount Wilson synoptic data for CRs 1815–1834. (b) A stackplot of the north polar regions above latitude N50°, constructed using the Mount Wilson synoptic data for CR 1815 and the simulated grid values for CRs 1816–1834. Here κ is 600 km^2 s^{-1} and v_0 is 10 m s^{-1}.

10.4 Simulations of the polar fields

In order to investigate this question, the simulation technique described in §9.6 was used, taking the MWO synoptic grid values for CR 1815 as the starting point and simulating the fields for CRs 1816–1834. In the simulation of the fields near minimum, each newly emerging active region is included in the grid of simulated fields by individual numerical substitution of the MWO values for the new region in the appropriate rotation. Near sunspot maximum, however, sunspots emerge frequently at lower latitudes, and it is not practical to carry out this procedure on an individual basis.

The numerical simulation is, instead, carried forward for one complete rotation, which consists of 60 integration steps, after which all grid values between latitudes $\pm 50°$ are updated by substituting the observed MWO values for this new rotation. The simulated values at higher latitudes are retained, however, and the simulations are continued for a further rotation, after which the procedure is repeated.

In this way one can be confident that the contributions from all emerging active regions with lifetimes of a solar rotation or more are included in the simulations of

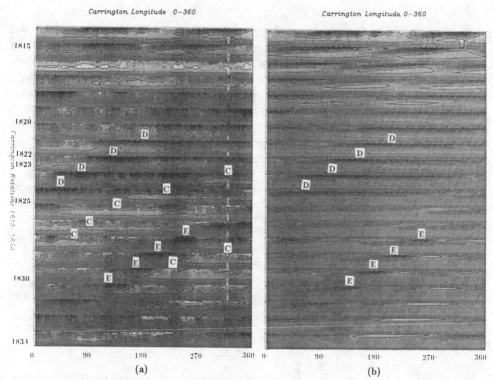

Fig. 10.4 As for Figure 10.3, showing the south polar regions below latitude S50°. The polar region lies at the bottom of each strip.

the polar fields. Further, possible saturation errors in the data, which may give rise to underestimates of the fields arising from decaying sunspot fields, are eliminated by using the actual solar field distribution at and below latitude N50° rather than the simulated values. In effect, this permits the Sun to do its own simulation of the large-scale field patterns below latitude N50° (thus eliminating problems related to the correct treatment of active-region nests and the possible subduction of flux loops) and invokes the flux transport equation only above this latitude.

In these calculations, the same models for the differential rotation and meridional flows were used, with $\kappa = 600$ km^2 s^{-1} and $v_0 = 10$ m s^{-1}. A stackplot of the simulated fields above latitude N50° for CRs 1816–1834 is shown in Figure 10.3(b), including the observed fields for CR 1815 for reference. A similar stackplot for the fields below latitude S50° is shown in Figure 10.4(b).

10.5 Discussion

It is obvious that, in the simulated stackplot, the north polar field is *not* reversed and that the penetration of the large-scale fields from lower latitudes, as predicted by the flux transport equation, fails to match the observed fields by a large margin. In the

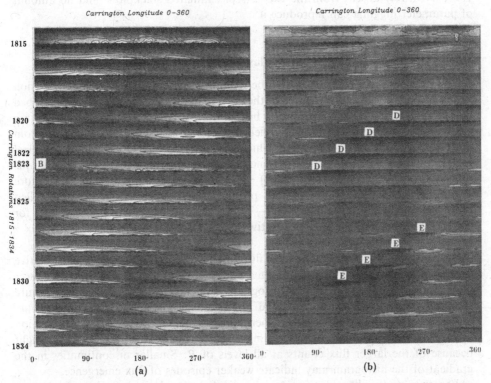

Carrington Longitude 0–360

Carrington Longitude 0–360

Fig. 10.5 (a) A stackplot of the north polar regions above latitude N50°, constructed using the Mount Wilson synoptic data for CR 1815 and the simulated grid values for CRs 1816–1834. Here κ is 1200 km^2 s^{-1} and v_0 is 50 m s^{-1}. (b) A stackplot of the south polar fields as for (a).

south, the simulated field regions are also much weaker than the observed regions. Other values of the parameters were tried, and stackplots of simulated fields, with $\kappa = 1200$ km^2 s^{-1} and $v_0 = 50$ m s^{-1}, are shown for both poles in Figure 10.5.

Some improvement in the agreement with the observed stackplots is obtained; the penetration of some of the flux regions is now closer to the observed patterns and, in the north, a 'reversal' has occurred by CR 1830. In the south, simulations with the larger parameter values also provide a better match for the negative regions marked D and E. It must be recognized, however, that the values of κ and v_0 used to produce Figure 10.5 are considerably larger than those based on observed values of large-scale velocity fields.

Despite these improvements, there are several significant qualitative discrepancies between the observed and simulated stackplots. In the north, the positive region marked A in Figure 10.3(a), which can be followed from CR 1822 through 1826, cannot be reproduced by any combination of parameters in the flux transport equation. Again, in the south (Figure 10.4(a)), an extended region of positive flux marked C, which can be seen in the observed stackplots from CR 1824 to CR

1829, is entirely absent from the two sets of simulated stackplots, and no amount of parameter juggling can reproduce it.

10.6 Flux histograms

A useful technique for distinguishing between flux pattern changes due to diffusion and meridional drift and those due to the emergence of new flux has been developed by Murray (1992). He has constructed histograms of the number of pixels $N(B)$ in a given area as a function of measured field strength B, based on magnetograms from the Kitt Peak station of the NSO, which consist of 2048×2048 one-arcsec pixel images of the solar disk. In the absence of newly emerging flux, such histograms follow a power law of the form, $N(B) \sim B^{-\alpha}$, and an example is shown in Figure 10.6(a), in which histograms of both positive (open squares) and negative (filled triangles) for the latitude range N60° − 70° are derived from magnetogram data for 23–31 December 1989. For both positive and negative flux, the power law index α is ~ 4.

Active regions, and other regions of flux emergence, produce excess counts relative to the power law. Strong new regions give rise to obvious departures from the power law at larger values of B in the histograms, and an example is shown in Figure 10.6(b). As an active region ages and breaks up, however, the more intense flux regions decay so that, if no new flux emerges in the region, the power law distribution is re-established, although the histogram curve may be uniformly shifted to the right because of the larger flux counts at all levels of B. Smaller discontinuities in the gradient of the histogram may indicate weaker episodes of flux emergence.

For this study, the solar image was divided into latitude strips ten degrees wide, and histograms were generated for each latitude zone by adding together a week's worth of magnetograms. Each histogram thus represents data from a region extending $\sim 90°$ in longitude and 10° in latitude. The emergence of flux in the form of active regions at latitudes below 40° is obvious, and examples are given by Murray (1992). Evidence of flux emergence was also found between latitudes N40° and N50° in the weeks beginning 20 November 1989 (Figure 10.6(b)), 16 January 1990 (Figure 10.6(c)) and 16 march 1990 (not shown). These may be identified with active regions in the appropriate NSO magnetograms and in the NOAA Boulder sunspot catalogue, where they are assigned the numbers 5804, 5886, and 5976, respectively. Smaller gradient discontinuities may also be seen in this latitude range on 2 November 1989, 8 January 1990, and at other times.

There is also evidence of flux emergence at higher latitudes. In the range N50° − 60°, for example, the histogram for the week beginning 23 December 1989, shown in Figure 10.6(d), exhibits characteristics similar to, but weaker than, those seen in Figure 10.6(c), where there is known to be an active region.

The slight excess of positive polarity field above 80 G in Figure 10.6(d) is not noise, for the source of this flux may be identified with a region of enhanced network near central meridian in the NSO magnetogram of 27 December (see Figure 10.6(b)). Reference back to Figure 10.3(a) shows that it may also be identified with the region marked A in that figure, since these regions correspond both in Carrington longitude (i.e. date of centre-disk passage) and latitude. Emerging flux signals were also noted

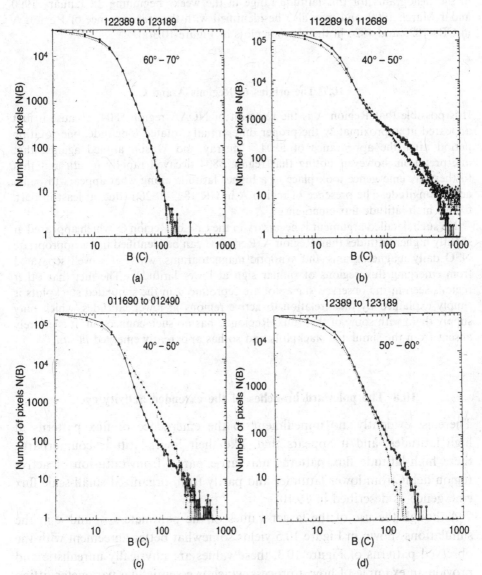

Fig. 10.6 Four sets of histograms of the number of pixels against field strengths for (a) latitude range N60°–70° for the week beginning 23 December 1989; (b) latitude N40°–50° for the week beginning 22 November 1989; (c) latitude N40°–50° for the week beginning 16 January 1990; (d) latitude N50°–60° for the week beginning 23 December 1989. The open rectangles represent positive flux and the filled triangles negative flux.

in the histograms for this latitude range in the weeks beginning 25 January 1990 and 1 March 1990, and may also be identified with the reappearance of Region A in CRs 1825 and 1826 in the contour plots of Figure 10.3(a).

10.7 The origins of Regions A and C

It is possible that Region A is the remnant of NOAA region 5804, because it first appeared at approximately the proper differentially rotated longitude, one rotation period after the appearance of 5804. Murray and Wilson argued against that interpretation, however, noting that region 5804 decayed rapidly *in situ* and that further flux emergence took place at a higher latitude along what appears to be an active longitude. The presence of region A in CR 1823–1826 is due, at least in part, to this high-latitude flux emergence.

No such detailed argument is necessary in the case of Region C, which appeared at slightly higher latitudes than Region A. Region C can be identified in the appropriate NSO daily magnetograms and synoptic magnetograms, where it is well separated from emerging flux regions of similar sign at lower latitudes. The fact that other regions seen in the observed stackplot are reproduced in the simulated stackplots is simply explained by their relation to active regions at lower latitudes, which may supply them with some of their flux. Region C has no such connection; it is entirely absent from the simulated stackplots and so has apparently emerged *in situ*.

10.8 The poleward branches of the extended activity cycle

There is, evidently, no impediment to the emergence of flux patterns at high latitudes, and it appears that, like their lower-latitude counterparts, these high-latitude flux patterns may arise partly from diffusion of active region fields from lower latitudes and partly from organized small-scale flux emergence as described in §9.10.

Although the use of the larger values of the parameters v_0 and κ in the simulations shown in Figure 10.5 yields somewhat better agreement with the observed patterns of Figure 10.3, these values are physically unrealistic and provide an example of how a process which is essentially a parameter-fitting exercise may conceal an important physical process.

Sometimes flux patterns may appear sequentially at both low and high latitudes along the same meridian of longitude, and the above discussion shows that the apparent poleward progress of these patterns may be due to sequential flux emergence along this meridian as part of a cell which extends over a comparable latitude range. Thus the concept of an active longitude may embrace a range of latitudes and both active regions and small-scale flux emergence. Other flux patterns may also emerge in the polar regions without connection to lower-latitude regions, Region C providing an unambiguous example.

In this discussion, the poleward branches of the EAC phenomena appear to play a significant role. The fact that these branches, whether derived from chromospheric green-line emission, torsional shear, or the polar crown, all reach the pole at the same time as the polar reversal can hardly be ignored. Further, the fact that all these features commence their poleward march at solar minimum and continue it *steadily* during the rising phase of the cycle is incompatible with the concept that this poleward march arises from the decay and poleward drift of active regions, which might otherwise be expected to be more sporadic.

10.9 The coronal holes and the polar reversals

The polar regions are the sites of the largest and longest-lived coronal holes at all times, except during the period just after sunspot maximum, when the polar magnetic fields are reversing. It has already been shown (Chapter 9) that the polar coronal hole extends to lower latitudes through the polar crown gap during the rising phase of the cycle and that, when the polar crown reaches latitudes $\sim 75°$, the gap closes, the polar crown moves rapidly polewards, and the reversal follows. Thus it is of interest to study the evolution of lower-latitude coronal holes during this period.

Cannon and Marquette (1990) have studied anomalously oriented active regions during the rising phase of Cycle 22. The magnetic axes of these regions have a positive inclination to the equator rather than the more usual negative inclination of $\sim 14°$. The leader flux of these regions is poleward of the follower flux and, in this configuration, would tend to maintain the existing polar fields, thus making a negative contribution to the polar field reversals according to the Leighton model. In all such cases, a lobe of the polar coronal hole (of the same polarity as the leader flux) extended down towards the leader flux and appeared to coalesce with it.

McIntosh (1992) has studied the equatorial coronal holes during the maximum and post-maximum phase of Cycle 21, and an example of the role of coronal holes in the polar field reversals can be seen in his synoptic charts. Just as the polar fields were beginning their reversals in 1979, a stable equatorial coronal hole appeared, and, during the following months, it was intermittently attached to the north polar coronal hole (of like polarity) via the polar crown gap. Significantly, the most outstanding active region of that year emerged adjacent to this coronal hole at the time of its maximum development. The polar crown gap closed after this event, the polar fields began the reversal process, and no further attachments to the polar hole occurred.

Later in the cycle (in 1982), the three most prolific X-ray flare regions of the cycle occurred: NOAA Regions 3776, 3763, and 4025. These three flare-producing regions emerged near large and rapidly growing coronal holes in their respective hemispheres. The coronal hole structure near Region 3763 is shown in the stackplots of the southern latitudes $1° - 50°$ in Figure 10.7. This plot shows the convergence of three separate coronal holes between CRs 1712 and 1721, and their coalescence, in CR 1722, which was accompanied by the rapid development of Region 3763 near this point. This region flared at the highest rate of the cycle and then rapidly fragmented, but the combined coronal holes became the largest coronal hole of the cycle, as can be seen in the compressed stackplot on the right of Figure 10.7.

The third region (Region 4025) formed near this large coronal hole during CR 1729, at a time when it began to diminish in area. During the following rotations a large element of this coronal hole structure formed north of the remnants of Region 4025 and migrated in a clockwise path around the active-region site, until it merged with the coronal hole lobe attached to the south pole. The development of this large southern coronal hole was coincident with the development of the largest of the northern coronal holes on the opposite side of the Sun and occurred at a time when the global magnetic patterns were changing to a simpler form. McIntosh suggests that these observations of coronal hole evolution may provide clues to the process mentioned in § 10.1, which appears to terminate the sunspot maximum phase and begin the gradual decline of the old cycle and the growth of the new polar coronal holes. This marks an important phase of the cycle and will be further discussed in Chapter 14.

While the evolution of equatorial coronal holes appears to be related to the polar coronal holes, that relationship, like the nature of the coronal hole itself, is far from understood. A coronal hole is identified with 'open' field lines extending outwards from the photosphere and, on account of its lateral and vertical extent, is unlikely to be a purely surface and atmospheric phenomenon. The subsurface configuration of coronal holes is, at this stage, a matter for pure speculation. Obridko and Shetling (1992) have proposed that they are indicators of the deep-seated global magnetic fields but do not provide any details of the field structures. The association of equatorial coronal holes with flaring active regions is intriguing, particularly since Hewish (1990) has argued that coronal holes, rather than the flaring regions, are the source of the particle streams which give rise to geomagnetic disturbances (see also § 14.7). In stackplots, such as those shown in Figure

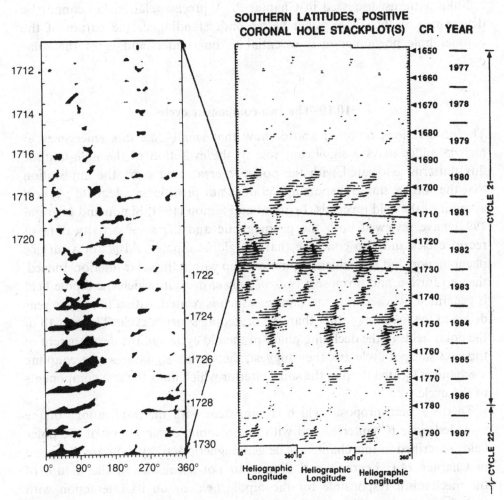

Fig. 10.7 Stackplots of HeI λ10830 coronal holes for the southern hemisphere latitude range 1° − 50°. At the right is a display of the southern-hemisphere coronal holes for the period 1977–1987 with the latitude scale compressed and the longitude range repeated three times in order to view the long-term drift of the patterns across heliographic longitude as in Figure 9.3(b). Horizontal lines in the middle of this display indicate the data interval shown at the left at expanded scale and with only a single range of 360° of Carrington longitude. (From McIntosh 1992)

10.7, they exhibit long-lived patterns similar to, but not identical with, the cell-like patterns discussed in Chapter 9. A precise relationship cannot be determined at this time, but a better understanding of the nature of the coronal hole phenomenon is essential to our understanding of the solar cycle.

10.10 The two-component cycle

The observations reported above show that small-scale flux emergence at high latitudes plays a significant role in the evolution of the high-latitude flux patterns associated with the polar reversals and carry the implication that the surface flux transport model does not provide an adequate *physical* account of the field reversals. Legrand and Simon (1991, Simon and Legrand 1992) have reviewed both the geomagnetic and the solar activity data of recent cycles and also conclude that the global dipole field is not a surface phenomenon, but one which originates deep within the solar interior. Indeed, they go further and propose a *two-component* cycle in which the dipole field is generated by a separate mechanism which is related, with a five to six year delay, to the *following* rather than the preceding sunspot cycle. The growth of the polar fields in the declining phase of the old cycle and the development of the *next* sunspot cycle are, they suggest, the surface signatures of a coupling mechanism located deep in the solar interior which links the two components of the cycle.

   This is a bold proposal which is consistent with the implications of the extended cycle (Chapter 8) and with the synoptic observations of the polar fields described in this chapter. (The geomagnetic activity data are discussed in Chapter 14.) Legrand and Simon do not speculate on the nature of the mechanism responsible for the dipole field or on its interaction with the sunspot-cycle mechanism, but various possibilities will be considered in subsequent chapters.

10.11 Conclusion

Although the kinematic models of the reversals of the polar fields have had some success, they lack a firm physical foundation, since they neglect the three-dimensional connectedness of magnetic fields. In this and the previous chapter, examples have been given in which such models do not correctly reproduce the observed evolution of the large-scale fields. Further, close inspection of small-scale fields fails to show a diffusion or spreading of the field. The small-scale flux has a lifetime of a few days at most, and the apparent spreading of a region is due to the regeneration of this flux over a slightly greater area. The unrealistically large values of the parameters required in the flux transport equation thus tend to conceal the underlying physical processes.

The alternative approach advocated by Stenflo (1992), by Legrand and Simon (see § 10.10) and by the author and his collaborators (Wilson, McIntosh, and Snodgrass 1990, Wilson and McIntosh 1991, Wilson 1992, and Murray and Wilson 1992) requires that the polar fields, and at least some of the low-latitude large-scale fields, arise from and are maintained by the *local* emergence of flux. Snodgrass and Wilson (1993, see § 9.10) have shown how the emergence of flux in the form of small bipoles may be organized (as required by Stenflo) by the sub-surface field and velocity configuration to produce the surface patterns independently of the decay of active regions.

The suggestion by Legrand and Simon of a two-component cycle in which the polar fields are generated by a mechanism separate from, but related to, the sunspot cycle mechanism, with a time delay of five to six years, fits comfortably with the phenomena of the extended cycle. This new approach offers hope of understanding the physical relation between the dipole and the active-region fields which is likely to differ significantly from that provided by the flux-transport model.

Clearly the mechanism responsible for the solar cycle is more complicated than any of the elementary models described in Chapter 6, and it is now appropriate to explore dynamo theory in more detail.

References

Babcock, H. W.: 1961, *Astrophys. J.*, **133**, 572.

Benevolenskaya, E. E., and Makarov, V. I.: 1992, in *Proceedings of the National Solar Observatory/Sacramento Peak 12th Summer Workshop*, ed. K. L. Harvey, San Francisco, California, 532.

Cannon, A., and Marquette, W.: 1990, in *Solar-Terrestrial Predictions Workshop* (Leura, Australia, 1989), NOAA, Boulder, Colorado, 331.

DeVore, C. R., and Sheeley, N. R.: 1987, *Solar Phys.*, **108**, 47–59.

Giovanelli. R. G.:, 1985, *Australian J. Phys.*, **38**, 1045.

Harvey, K. L.: 1992, in *Proceedings of the National Solar Observatory/Sacramento Peak 12th Summer Workshop*, ed. K. L. Harvey, San Francisco, California, 335.

Hewish, A.: 1990, in *Solar-Terrestrial Predictions Workshop* (Leura, Australia, 1989), NOAA, Boulder, Colorado, 81.

Legrand, J. P., and Simon, P. A.: 1991, *Solar Phys.*, **131**, 187.

Leighton, R. B.: 1964, *Astrophys. J.*, **140**, 1547.

McIntosh, P. S.: 1992, in *Proceedings of the National Solar Observatory/Sacramento Peak 12th Summer Workshop*, ed. K. L. Harvey, San Francisco, California, 14.

Murray, N.: 1992, *Astrophys. J.*, (in press).

Murray, N., and Wilson, P. R.: 1992, *Solar Phys.,* **142**, 221.

Obridko, V., and Shetling, B.: 1992, in *Proceedings of the National Solar Observatory/Sacramento Peak 12th Summer Workshop*, ed. K. L. Harvey, San Francisco, California, 187.

Simon, P. A., and Legrand, J. P.: 1992, *Solar Phys.,* **141**, 391.

Snodgrass, H. B., and Wilson, P. R.: 1993, *Solar Phys.,* (in press).

Stenflo, J. O.: 1992, in *Proceedings of the National Solar Observatory/Sacramento Peak 12th Summer Workshop*, ed. K. L. Harvey, San Francisco, California, 83.

Wilson, P. R., McIntosh, P. S., and Snodgrass, H. B.: 1990, *Solar Phys.,* **127**, 1–9.

Wilson, P. R., and McIntosh, P. S.: 1991, *Solar Phys.,* **136**, 221–237.

Wilson, P. R.: 1992, *Solar Phys.,* **138**, 11–21.

11
The role of dynamo theory in cyclic activity

11.1 Introduction

The recognition that magnetic fields are an essential component not only of solar and stellar activity but also of the structure of galaxies, quasars, and pulsars has focussed considerable theoretical interest on the origin and maintenance of cosmic magnetic fields. Since the length scales associated with many cosmic magnetic fields are very large, the ohmic decay times (see §4.1 and below) are long, and there is no difficulty in explaining the continued existence of primordial or fossil fields, such as the megagauss fields found in magnetic A-type stars; but the changes observed to occur in many cosmic magnetic fields, over periods which may be short compared with the decay time, entail an interaction between the plasma motions and the existing fields which may also maintain these fields against ohmic decay. This has become known as *dynamo action*, and, in order to understand evolutionary changes occurring in the solar magnetic cycle, it is necessary to probe further into the underlying theory.

Parker (1970) drew attention to the curious asymmetry throughout the universe between electric and magnetic charge on the one hand, and the corresponding fields on the other. The abundant free electric charges rapidly neutralize electric fields in the local frame of the gas, but the apparent non-existence of magnetic monopoles means that, whenever there is a net relative motion of electrons and ions in a gas, implying a current density $\mathbf{j}$, Maxwell's equation (see Equation 4.2),

$$\frac{\mathbf{j}}{\mu} = \nabla \times \mathbf{B}, \tag{11.1}$$

(where μ is the magnetic permeability) requires the presence of an inhomogeneous magnetic field $\mathbf{B}$.

The temporal evolution of magnetic fields is described by the hydromagnetic induction equation (assuming uniform magnetic diffusivity η)

$$\frac{\partial \mathbf{B}}{\partial t} = \nabla \times (\mathbf{u} \times \mathbf{B}) + \eta \, \nabla^2 \, \mathbf{B}, \tag{11.2}$$

which was derived as Equation (4.9) in Chapter 4. It follows that, in a stationary medium ($\mathbf{u} = \mathbf{0}$), there is no source term and the field can only decay. The 'ohmic' decay time is $\tau = L^2/\eta$, where L is the scale length of the system. For the Sun's global field, Cowling (1976) has estimated a decay time $\tau_\odot \approx 10^{10}$ years, while for sunspots his estimate is ~ 300 years. In other astrophysical situations, η is also expected to be small. In the limit $\eta = 0$, $\mathbf{u} \neq \mathbf{0}$, the field moves with the fluid (see §4.1) and is said to be frozen-in. Dynamo action, which involves changes in the field relative to the plasma over periods less than the decay time, depends on a delicate interaction between the two terms on the right hand side of Equation (11.2).

The plasma velocity, $\mathbf{u}$, is, of course, governed by the momentum equation which includes the Lorentz term, $\mathbf{j} \times \mathbf{B}$ (see Equation 4.15), but in kinematic dynamo theory one assumes that $\mathbf{u}$ may be prescribed independently of $\mathbf{B}$. For a given $\mathbf{u}$, the possibility of dynamo action can be assessed by seeking solutions to Equation (11.2) of the form $\mathbf{B} = \mathbf{B}_0(\mathbf{r}) \exp(qt)$ and determining the eigenvalues of q. If $q = 0$, the field is steady, while if q is purely imaginary, the field is oscillatory or wavelike, but, in either case, one has a *neutral* dynamo (i.e. the field neither grows nor decays exponentially). In general q is complex, and, for sufficiently small $\mathbf{u}$, q must have a negative real part (since, for $\mathbf{u} = 0$, the field must decay), so the aim is to find a pattern of motion such that, as $\mathbf{u}$ increases, $\mathrm{Re}(q) \geq 0$. Such a pattern of motion is said to provide *dynamo maintenance* of the field.

From this basic concept, dynamo theory has blossomed into an independent discipline in which many dynamo theorems and anti-dynamo theorems have been stated (and sometimes proven). Dynamo maintenance has been established for a wide range of steady motions and the existence of 'fast dynamos', i.e. those that can function in the limit of zero diffusivity, has been postulated but not proven. The interested reader is referred to any of the many excellent reviews of the field: e.g. Roberts (1971), Moffatt (1974, 1976), Cowling (1981), Rosner and Weiss (1992). Here we summarize those aspects of dynamo theory which are likely to be most relevant for solar and stellar cycles.

11.2 The dynamo equations

Since a cartesian system contains all the physics of the problem and avoids the mathematical complications of a spherical polar geometry, we first postulate a cartesian system but assume that it is located at a co-latitude angle θ in the northern hemisphere. The z-axis is in the outward radial direction, the x-axis is in the meridional plane directed towards the equator, and the y-axis lies in the azimuthal plane in the direction of rotation.

Assuming that the velocity $\mathbf{u}$ is given, the kinematic dynamo is governed by Equation (11.2); for an axially symmetric magnetic field, the poloidal component is defined in terms of the vector potential $\mathbf{A} = A(x, z, t)\mathbf{j}$ and the azimuthal or toroidal component by $B(x, z, t)$. Thus

$$\mathbf{B}(x, z, t) = -\frac{\partial A}{\partial z}\mathbf{i} + B\mathbf{j} + \frac{\partial A}{\partial x}\mathbf{k} = B\mathbf{j} + \nabla \times \mathbf{A}. \tag{11.3}$$

The velocity field may be defined by

$$\mathbf{u} = u(x, z, t)\mathbf{j} + \mathbf{v}_c(x, y, z, t),$$

where $u\mathbf{j}$ represents the rotational velocity and $\mathbf{v}_c$ the convective motion in the $x - z$ plane.

Substitution in equation (11.2) yields

$$\nabla \times \left\{ \left(\frac{\partial}{\partial t} - \eta\nabla^2 \right) A\mathbf{j} \right\} + \left(\frac{\partial}{\partial t} - \eta\nabla^2 \right) B\mathbf{j}$$
$$= \{\nabla u \times \nabla A\} + \nabla \times \{\mathbf{v}_c \times (B\mathbf{j} + \nabla \times A\mathbf{j})\}. \tag{11.4}$$

Taking the $\mathbf{j}$-component yields

$$\left(\frac{\partial}{\partial t} - \eta\nabla^2 \right) B + \mathbf{v}_c \cdot \nabla B = \{\nabla u \times \nabla A\} \cdot \mathbf{j} - B \nabla \cdot \mathbf{v}_c, \tag{11.5}$$

and, after integrating (or 'uncurling') the remainder of Equation (11.4),

$$\left(\frac{\partial}{\partial t} - \eta\nabla^2 \right) A + \mathbf{v}_c \cdot \nabla A = (\mathbf{v}_c \times B\mathbf{j}) \cdot \mathbf{j}. \tag{11.6}$$

The $\mathbf{v}_c \cdot$-terms on the left sides of Equations (11.5) and (11.6) simply describe the advection of the field and do not contribute to dynamo action. The term $\{\nabla u \times \nabla A\} \cdot \mathbf{j}$ in Equation (11.5) is a 'source term', whereby the radial and meridional gradients in the rotational velocity act on the poloidal field to generate the toroidal field, an action called the *ω-effect*. Unfortunately, the equivalent term on the right side of Equation (11.6), $\mathbf{v}_c \times (B\mathbf{j}) \cdot \mathbf{j}$, is identically zero for arbitrary $\mathbf{v}_c$, and thus B cannot make a similar contribution to the maintenance of the poloidal field, which consequently must decay. This is the essence of Cowling's theorem, which entails that a strictly axisymmetric field cannot be maintained, even by non-axisymmetric motions.

Following a physical discussion of the rate of production of the poloidal field by convective motions, however, Parker replaced the right-hand side of Equation (11.6) by ΓB, where Γ is a parametric representation of the effects of cyclonic convection, i.e.

$$\left(\frac{\partial}{\partial t} - \eta \nabla^2\right) A = \Gamma B. \tag{11.7}$$

Parker's physical picture of Ω-loops forming in the toroidal field, then rising and twisting under the combined effects of buoyancy and the Coriolis forces (§6.8), clearly entails the creation of non-axisymmetric fields as part of the process of generating poloidal fields from toroidal fields. However, his daring mathematical legerdemain in writing Equation (11.7) in place of (11.6) left many people breathless, and a somewhat more rigorous formulation seemed desirable.

11.3 Mean-field electrodynamics

The essence of the problem is to describe in mathematical terms how the global solar field, which is crudely axisymmetric at any epoch of the cycle, yet displays large non-axisymmetric features (active regions and, particularly, active longitudes) with intensities perhaps two orders of magnitude greater than the axisymmetric field, can undergo regular periodic interchange between its poloidal and toroidal components, a process in which the larger non-axisymmetric fields appear to be causally involved.

The best-known attempt to address this problem is by Steenbeck, Krause, and Rädler and has become known as *mean-field electrodynamics* (MFE) (Krause and Rädler 1980). The velocity $\mathbf{u}$ and the magnetic field $\mathbf{B}$ are expressed as

$$\mathbf{u} = \langle \mathbf{u} \rangle + \mathbf{u}', \quad \mathbf{B} = \langle \mathbf{B} \rangle + \mathbf{B}',$$

where the terms $\langle \mathbf{u} \rangle$ and $\langle \mathbf{B} \rangle$ represent the slowly varying mean components and $\mathbf{u}'$ and $\mathbf{B}'$ the non-axisymmetric fluctuating components. The turbulent motion $\mathbf{u}'$ is assumed to possess a correlation time τ and correlation length λ, which are small compared to the scale time t_0 and scale length l_0 of the variations of $\langle \mathbf{u} \rangle$ and $\langle \mathbf{B} \rangle$. Thus τ is a mean time, after which $\mathbf{u}'$ is no longer correlated with its initial value, and λ is comparable with the mean eddy size. The operator $\langle \rangle$ implies an azimuthal average, and it is assumed that $\langle \mathbf{u}' \rangle$ and $\langle \mathbf{B}' \rangle$ are zero.

Substitution in Equation (11.2) and application of the operator $\langle \rangle$ yields

$$\frac{\partial \langle \mathbf{B} \rangle}{\partial t} = \nabla \times (\mathbf{E} + \langle \mathbf{u} \rangle \times \langle \mathbf{B} \rangle) - \nabla \times (\eta \nabla \times \langle \mathbf{B} \rangle) \tag{11.8}$$

and, subtracting this from the complete equation,

$$\frac{\partial \mathbf{B}'}{\partial t} = \nabla \times (\langle \mathbf{u} \rangle \times \mathbf{B}' + \mathbf{u}' \times \langle \mathbf{B} \rangle + \mathbf{G}) - \nabla \times (\eta \nabla \times \mathbf{B}'), \tag{11.9}$$

where

$$\mathbf{E} = \langle \mathbf{u}' \times \mathbf{B}' \rangle, \quad \mathbf{G} = \mathbf{u}' \times \mathbf{B}' - \langle \mathbf{u}' \times \mathbf{B}' \rangle.$$

In Equation (11.8), $\mathbf{E}$ may be regarded as a mean electric field arising from the interaction of the turbulent motion and the magnetic field, to be determined by solving Equation (11.9) for $\mathbf{B}'$.

In order to do this, various assumptions are necessary. The assumption that $\langle \mathbf{u}' \rangle = \mathbf{0}$ may be not unreasonable for a fully turbulent velocity field but may be questionable for a convective field which is sufficiently ordered for Coriolis effects to be important. The 'first-order-smoothing' approximation, $\mathbf{G} \approx \mathbf{0}$, which is valid only if $\mathbf{B}'$ is small compared to $\langle \mathbf{B} \rangle$, is, for mathematical convenience, also used in Equation (11.9), despite the fact that it obviously does not apply to the solar fields. Equation (11.9) thus becomes

$$\frac{\partial \mathbf{B}'}{\partial t} + \nabla \times (\eta \, \nabla \times \mathbf{B}') = \nabla \times (\mathbf{u}' \times \langle \mathbf{B} \rangle). \tag{11.10}$$

In determining $\mathbf{E}$, only $\mathbf{B}''$, the component of $\mathbf{B}'$ which is correlated with $\mathbf{u}'$, need be considered and, by the definition of τ, $\mathbf{B}(t + \tau)$ is not correlated with $\mathbf{B}(t)$ for any t. Then $\mathbf{B}''(t)$ may be determined by integrating Equation (11.10) from $t - \tau$ to t, with the initial condition $\mathbf{B}'(t - \tau) = \mathbf{0}$. Further, since τ is, at most, of the order of the eddy turnover time λ/v, both $\mathbf{u}'$ and $\langle \mathbf{B} \rangle$ may be regarded as independent of t, and the integration of Equation (11.10) becomes elementary. The right-hand side is linear in $\langle \mathbf{B} \rangle$ and its derivatives; therefore, using the summation convention,

$$E_i = \alpha_{ij} \langle B_j \rangle + \beta_{ijk} \frac{\partial \langle B_j \rangle}{\partial x_k}, \tag{11.11}$$

where α_{ij} and β_{ijk} depend on the local structure of the velocity field and on τ. If the turbulent field is isotropic, $\alpha_{ij} = \alpha \delta_{ij}$, $\beta_{ijk} = \beta \epsilon_{ijk}$, and

$$\mathbf{E} = \alpha \langle \mathbf{B} \rangle - \beta \, \nabla \times \langle \mathbf{B} \rangle. \tag{11.12}$$

If τ is small compared to the decay time λ^2/η, the diffusive term may be neglected in Equation (11.10), leading to

$$\alpha = -\frac{1}{3} \tau \langle \mathbf{u}' \cdot \nabla \times \mathbf{u}' \rangle, \quad \beta = \frac{1}{3} \tau v^2. \tag{11.13}$$

Thus, dropping the brackets, Equation (11.8) for the slowly varying axisymmetric field becomes

$$\frac{\partial \mathbf{B}}{\partial t} = \nabla \times (\alpha \mathbf{B} + \mathbf{u} \times \mathbf{B}) - \nabla \times \{(\eta + \beta) \, \nabla \times \mathbf{B}\} \tag{11.14}$$

This is similar to Equation (11.2) but contains the term $\alpha \mathbf{B}$, which represents an electric field parallel to the magnetic field and enables the theory to escape from Cowling's theorem. Equation (11.14) also contains an eddy-diffusivity coefficient, β, which operates by mixing magnetic fields transported from neighbouring regions. It does not destroy the resulting small-scale field but may reduce it to a scale at which it can be eliminated by ohmic diffusion (some difficulties with this concept are discussed in § 11.5).

In the mean-field dynamo, the magnetic diffusivity η is replaced by a total diffusivity η' $(= \eta + \beta)$, and Equation (11.2) becomes

$$\frac{\partial \mathbf{B}}{\partial t} = \nabla \times (\alpha \mathbf{B} + \mathbf{u} \times \mathbf{B}) + \eta' \nabla^2 \mathbf{B}. \tag{11.15}$$

In what follows, it is customary to drop the prime on η, but it should be remembered that the presence of the α term implies that η is the turbulent diffusivity. Since $\mathbf{B}$ is axisymmetric here, it may be represented by the poloidal and toroidal terms, $A(x, z, t)$ and $B(x, z, t)$, as in Equation (11.3), while $\mathbf{u} = u(x, z, t)\mathbf{j}$. If the advection terms are neglected, Equation (11.15) becomes

$$\left(\frac{\partial}{\partial t} - \eta \nabla^2 \right) B = \{\nabla u \times \nabla A\} \cdot \mathbf{j} - \alpha \nabla^2 A \tag{11.16}$$

and

$$\left(\frac{\partial}{\partial t} - \eta \nabla^2 \right) A = \alpha B. \tag{11.17}$$

Unlike Equation (11.6), Equation (11.17) contains a non–zero source term, αB on the right-hand side, and dynamo action is possible since Equations (11.16) and (11.17) provide for the regeneration of both the toroidal and the poloidal fields.

Of the two source terms in Equation (11.16), that arising from non-uniform rotation, described by ∇u, is likely to be larger than that involving α, and, if the latter is neglected, Equations (11.16) and (11.17) constitute what has become known as the α–ω dynamo. Equation (11.16) describes the ω-effect, whereby the poloidal field is sheared by non-uniform rotation to generate the toroidal field, while the essential feedback, or α-*effect*, is provided by the *helicity* $\mathbf{v}_c \cdot \nabla \times \mathbf{v}_c$ of the non-axisymmetric cyclonic convection, which generates an azimuthal electromotive force $\mathbf{E}$, proportional both to the helicity and to B_ϕ, leading to Equation (11.17). Mathematically Equations (11.7) and (11.17) are equivalent, but η is now the turbulent diffusivity, and the parameter α, which replaces Γ, is directly proportional to the helicity of the flow, although of opposite sign.

Defining a characteristic length scale ℓ_0, a decay time $t_0 = \ell_0^2/\eta$, and $u = s_0\omega$, where s_0 is of the order of the local radius of rotation and ω is the local angular velocity, Equations (11.16) and (11.17) may be written in terms of the non-dimensional variables $t' = t/t_0$ and $\mathbf{r}' = \mathbf{r}/\ell_0$. Eliminating B and neglecting the α^2 term yields

$$\left(\frac{\partial}{\partial t'} - \nabla'^2 \right)^2 A = \frac{\alpha \ell_0^2 s_0}{\eta^2} \{\nabla'\omega \times \nabla'A\} \cdot \mathbf{j}. \tag{11.18}$$

If α_0 and ω_0 are scale factors giving the orders of magnitude of α and $|\nabla'\omega|$, Equation (11.18) becomes

$$\left(\frac{\partial}{\partial t'} - \nabla'^2 \right)^2 A = D \frac{\alpha}{\alpha_0} \left\{ \frac{\nabla'\omega}{\omega_0} \times \nabla'A \right\} \cdot \mathbf{j}, \tag{11.19}$$

where the non-dimensional *dynamo number D* is defined by

$$D = \frac{\alpha_0 \omega_0 \ell_0^2 s_0}{2\eta^2}. \tag{11.20}$$

Dynamo action might be expected when D exceeds some critical value.

Dynamo action is also possible in situations where ∇u is negligible compared to α, and only the second term is retained in Equation (11.16). Such dynamos are called α^2 dynamos, since both the source terms in Equations (11.16) and (11.17) are proportional to α. If both terms on the right of Equation (11.16) are comparable, the dynamo is said to be an $\alpha^2\omega$ dynamo.

For solar-type stars with well-structured convection zones, the $\alpha-\omega$ dynamo would be the likely dynamo mode. However, other dynamo modes may operate in other types of stars. Gray (1992) has noted the sudden development of activity in the evolution of luminosity class III giant stars as they cross the 'granulation boundary' and develop a convective envelope (see §7.6), and the subsequent discontinuous decrease in rotation rate. These stars have shallow convection zones and, consequently, differential rotation in latitude is likely to be small. Further, during the evolutionary phase before rotational braking occurs, there is nothing to generate a radial angular velocity gradient. Without velocity gradients in some direction, $\nabla u = 0$ in Equation (11.16) and the $\alpha-\omega$ dynamo cannot function. Gray argues that the activity which develops during this phase is evidence for the development of an α^2 dynamo.

As a star approaches the rotation discontinuity boundary, however, the effectiveness of the α^2 dynamo may increase rapidly, because it is sensitive to the increasing depth of the convection zone which, at this stage, is $\sim 6\%$ of the radius. Gray argues that, as the magnetic field grows, so does the magnetic braking, and the convective envelope suffers shear stress resulting in strong radial gradients, especially at the interface between the convection zone and the radiative core below it. The $\alpha-\omega$ dynamo is now switched on, resulting in strong magnetic braking.

This argument appears plausible, but it is unlikely that such dynamos are applicable to the fields of solar-type stars with their well-developed convection zones for which an $\alpha-\omega$ dynamo is the most likely mechanism. We now discuss application of the $\alpha-\omega$ dynamo to the concept of the migratory dynamo wave.

11.4 The migratory dynamo

If an α–ω dynamo is responsible for the magnetic changes associated with the solar cycle, then it seems likely that it must take the form of a dynamo wave, as plausibly suggested by Parker (1955) and described heuristically in §6.8. Wave phenomena, however, must be expressible in terms of a wave equation, from which a dispersion relation for the wave growth rate (e.g. q above) may be derived, and hence the phase speed deduced and compared with the observed phenomena.

In Equation (11.3), consider wavelike solutions of the form

$$A = A_0 \exp\{i(k_1 x + k_3 z - vt)\}, \quad B = B_0 \exp\{i(k_1 x + k_3 z - vt)\}, \qquad (11.21)$$

where the wave vector $\mathbf{k}^* = k_1 \mathbf{i} + k_3 \mathbf{k}$ defines the direction and sense of the wave propagation, provided that the real part of v is positive. However, if v is complex, the imaginary part of v determines whether the wave will grow or decay.

The substitution of Equation (11.21) into Equations (11.16) and (11.17) yields the dispersion relation

$$v = -i\eta(k^*)^2 \pm \{\alpha(\mathbf{k}^* \times \nabla u) \cdot \mathbf{j}\}^{\frac{1}{2}} \frac{(1+i)}{2^{\frac{1}{2}}}. \qquad (11.22)$$

If the wave is to grow in amplitude, the imaginary part of v must be positive. The term $-i\eta(k^*)^2$ is negative, however, so the positive sign must be chosen in Equation (11.22) (whether $\{\alpha(\mathbf{k}^* \times \nabla u) \cdot \mathbf{j}\}^{\frac{1}{2}}$ is real or imaginary), and $|\{\alpha(\mathbf{k}^* \times \nabla u) \cdot \mathbf{j}\}^{\frac{1}{2}}|$ must exceed $2^{\frac{1}{2}}\eta(k^*)^2$. The real part of v must also be positive (since $\mathbf{k}^*$ defines the sense of propagation), and, if α is positive (i.e. the helicity is negative), as is expected in the Sun, $(\mathbf{k}^* \times \nabla u) \cdot \mathbf{j}$ must be positive. Maximizing the wave's growth rate implies that the propagation direction $\mathbf{k}^*$, the direction in which u increases most rapidly (i.e. the direction of ∇u), and the direction of rotation $\mathbf{j}$ must be mutually perpendicular and form a right-handed set. If α is negative (i.e. the helicity is positive), they form a left-handed set. This result is consistent with Yoshimura's (1976) finding that dynamo waves propagate along iso-rotation surfaces.

The condition that

$$|\{\alpha(\mathbf{k}^* \times \nabla u) \cdot \mathbf{j}\}^{\frac{1}{2}}| > 2^{\frac{1}{2}}\eta(k^*)^2$$

may be written in terms of the standardized quantities defined above, i.e.

$$\frac{\alpha_0 \omega_0 \ell_0^2 s_0}{2\eta^2} > 1,$$

Thus the dynamo number D must exceed 1.

The direction of propagation of a solar dynamo wave may therefore be assessed in terms of the configuration of the iso-rotation surfaces within the Sun. In the 1980s, Gilman (1983), Gilman and Miller (1986), Glatzmaier (1985), and others, carried out extensive non-linear dynamical calculations of the internal rotation profile of the Sun. These indicated that the angular velocity of rotation should increase outwards through the convection zone, the iso-rotation surfaces being (approximately) cylinders of constant radius about the rotation axis. If positive helicity is assumed, the right-hand rule indicates that a dynamo located within the convection zone should propagate polewards, contrary to the evidence of the butterfly diagram, a result which Parker (1987a) has characterized as the 'dynamo dilemma'.

Recent findings from helioseismology have shown that this rotation model is inconsistent with the observed splittings of the p-mode frequencies. Rather than resolving the dynamo dilemma, however, these results have raised other problems for the dynamo wave model. Many variations of the kinematic $\alpha-\omega$ dynamo have been put forward in order to resolve this and other problems for the solar dynamo, and some of these are discussed in the next chapter.

11.5 Other problems of solar dynamo theory

While the development of MFE provides mathematical support for Parker's bold conjecture, which led to Equation (11.7), it is strictly applicable only to situations for which its assumptions are valid, in particular, that $\mathbf{B}'$ is small compared to $\langle \mathbf{B} \rangle$. Although this assumption may be applicable to the magnetic field of the earth, other planets, and perhaps some stars, it does not apply to the solar field, where $\mathbf{B}'$ must include the fields associated with active regions, which are comparable with the toroidal fields and greater than the azimuthally averaged poloidal fields by several orders of magnitude. Despite its greater mathematical rigour, MFE would therefore seem to have no greater validity for application to the Sun than Parker's heuristic arguments to justify Equation (11.7).

Another problem is the apparent imbalance between the magnitude of the poloidal fields (a few gauss at the poles) and the toroidal fields (several kilogauss), although they are regarded as comparable in Equations (11.16) and (11.17). At best, the system of Equations (11.16) and (11.17) provides a reliable parametrized theory which captures the essential physics of turbulent dynamo action and forms the basis of much discussion of the stellar application of dynamo theory.

There are other problems, however, in applying mean-field dynamo theory to the Sun. The Sun's magnetic field cannot be regarded simply as being passively advected by plasma motions, and this calls into question the basic kinematic assumption underlying the mean-field dynamo. The removal of this assumption is discussed in §11.6.

The problem of magnetic buoyancy, which must also be considered, raises the question of the actual location of the dynamo process. It is well known (see §4.5) that magnetic fields are buoyant in a stratified medium, but magnetic buoyancy is not included in the mean-field equations. Piddington (1972) argued that magnetic buoyancy would lead to a progressive expulsion of flux from the convection zone, such that eventually all magnetic flux would leave the solar interior, thus terminating all dynamo action. Parker (1975) showed that for mean field strengths > 100 G the buoyancy rise rate would be sufficient to suppress the α-effect, a result which suggests that the dynamo should lie deep within the convection zone. Parker (1987b) has argued more recently that a toroidal subsurface magnetic field may survive the effects of buoyancy through the development of a 'thermal shadow' above the field, owing to the reduced efficiency of convection across the field. This would make the shadow cooler and denser than its surroundings, and Parker calculates that the net effect should counteract the buoyancy of the field.

Leighton (1969) attempted to parametrize the buoyancy effect, but such attempts are unsatisfactory because they may not properly describe the buoyancy process in a convectively unstable fluid, such as the convection zone. Concerns about flux storage over the 11-year cycle led Golub *et al.* (1981), and others, to suggest that the dynamo was not located in the convection zone, but rather in the undershoot region which separates the convectively unstable layers from the stably stratified radiative interior, and here magnetic buoyancy can be treated as an instability problem. How flux generation would occur in such a layer or whether the calculations would give a reasonable account of the magnetic structures which ultimately form is far from clear.

Another important problem for the MFE dynamo is its treatment of the diffusion term. It has long been recognized that solar magnetic fields change on time-scales which are incompatible with the normal molecular diffusivity, η, and, in the MFE formulation, this is replaced by a turbulent diffusivity, $\eta' = \eta + \beta$, where β (see Equation 11.13) describes the turbulent velocity field. It is envisaged that the field is dispersed as a result of the mixing of adjacent magnetic fields by turbulent velocity fields. Parker likened the process to the turbulent diffusion of smoke, but it is not clear that such a process is applicable to a vector field.

Indeed, Piddington (1972) argued vigorously against this view, pointing out that small-scale magnetic fields in a turbulent velocity field may grow by stretching and twisting faster than they can be dissipated, and that the only limitation on this growth would be the back-reaction of the Lorentz force. Instead of dissipation, Piddington envisaged a final state in which the small-scale fluid motions are ultimately damped and give rise to motions corresponding to a superposition of Alfvén waves.

This problem has been re-examined by Rosner and Weiss (1992), who found that, provided the field remains 'kinematic' (i.e. the Lorentz back-reaction may be ignored), turbulent diffusion 'works'. However, only for large-scale initial fields, such that

$$|B_0| < R_{\mathrm{m}}^{-\frac{1}{2}} B_{\mathrm{e}}, \tag{11.23}$$

where $R_{\mathrm{m}} = u\ell/\eta$ is the magnetic Reynolds number(u and ℓ being typical velocity and length scales) and B_{e} is the equipartition field, will the system remain kinematic. Unfortunately, for values of R_{m} appropriate to the Sun, this critical field strength is several orders of magnitude less than the observed large-scale solar fields.

Cattaneo and Vainshtein (1991) have carried out two-dimensional numerical simulations to study the process of turbulent diffusion and found that, initially, the magnetic field grows over a wide range of field strengths. Provided the overall field is weaker than the limit given by Equation (11.23), however, this growth is rapidly overwhelmed by a cascade of magnetic energy to smaller scales, where it is thermalized on a resistive time-scale. For fields in excess of the critical value, however, the cascade process is replaced by an apparent stasis in which the energy in the large-scale field components changes only very slowly. This gradual decay continues until the strength of the large-scale field has decreased to a level at which the cascade process can resume. Regrettably for the mean field-dynamo, Cattaneo and Vainshtein find that, for $B_{\mathrm{e}}=500$ G and $R_{\mathrm{m}} = 10^{12}$, this critical field strength is only 5×10^{-4} G!

11.6 Non-linear dynamos

In the kinematic dynamo the magnetic energy can grow without limit but, in reality, the growth of the field must be limited by the Lorentz force which, according to the equation of motion (Equation 4.17), must modify the velocity field. By the construction of simplified models, several effects of the Lorentz force can be isolated: it may suppress helicity and so quench the α-effect; it may also modify the differential rotation, a process which can be represented by the equation

$$\frac{\partial \mathbf{u}}{\partial t} = \frac{1}{\mu_0 \rho} \{ (\nabla \times \mathbf{B}_T) \times \mathbf{B}_P \} \mathbf{j} + \nu \nabla^2 \mathbf{u}, \tag{11.24}$$

where $\mathbf{u}$ is the toroidal velocity perturbation and ν is the turbulent viscous diffusivity. Since the Lorentz force is quadratic in the magnetic field, $\mathbf{u}$ can have both a steady component and a component with twice the frequency of the dynamo (Kleeorin and Ruzmaikin 1991). If this process is appropriate for a solar dynamo, the period would be 11 years. It has been suggested (e.g. Yoshimura, 1981) that such perturbations might be identified with torsional oscillations.

Qualitative features of non-linear dynamo systems have been inferred from truncated models consisting of low-order systems of ordinary differential equations. These elementary systems exhibit patterns of behaviour which are shared by solutions of the full partial differential equations. In his study of the one-dimensional dynamo wave, Parker (1979) showed that there is an oscillatory (Hopf) bifurcation, leading to exponentially growing solutions of the kinematic problem when D is of order 1. Non-linear extensions of this problem, with growth limited by various saturation processes, can be used to illustrate the development of complex time-dependent behaviour. Periodic non-linear solutions can be constructed in cases where the dynamo number $D > 1$ (Weiss, Cattaneo, and Jones 1984), while chaotic oscillations have been obtained for $D > 4$ (Jones, Weiss, and Cattaneo 1985). Periodic solutions exhibit two time-scales: one corresponding to the magnetic cycle, the other corresponding to irregular modulations on a longer time-scale, showing that grand minima can appear naturally even in a very simple model.

While solutions of the kinematic problem have either dipole or quadrupole symmetry, with eigenfunctions whose toroidal components are either symmetric or antisymmetric about the equator, these symmetries may be broken in the non-linear domain at secondary bifurcations which lead to the appearance of asymmetric oscillations resembling the observed behaviour of the Sun. These models also display another important feature of non-linear systems: the same set of parameter values may yield several different stable solutions as well as many unstable solutions.

The next stage is to construct mean-field dynamos containing some arbitrary non-linear quenching mechanism. The simplest models are one-dimensional (e.g. Stix 1989, Moss *et al.* 1990), and these also provide examples of symmetry breaking and complicated time dependence. There have been many axisymmetric dynamo models, with both radial and latitudinal variations of the field, which, when suitably tuned, produce dynamo waves that migrate either towards the equator or towards the pole (e.g. Parker 1979, Stix 1989). By modifying the spatial variations of α or ω, it is possible to generate simultaneous poleward and equatorward migrations (Stix 1987, D. Schmitt 1987). Mean-field dynamos also provide a means of exploring the effects of different quenching mechanisms (Noyes *et al.* 1984) or the relative importance of the α- and the ω-effects in $\alpha^2\omega$ dynamos (Jennings, 1992). In most non-linear dynamo models, the cycle frequency increases with increasing dynamo number, as inferred from observation.

On a grander scale, Gilman (1983) and Glatzmaier (1985) have attempted numerical solutions of the full partial differential equations, including the equation of motion, without averaging over small-scale eddies. Gilman obtained a dynamo driven by convection in a Boussinesq fluid contained within a rotating annulus with dynamo waves propagating towards the pole. Glatzmaier obtained similar results in the anelastic approximation. (In the anelastic approximation, the generation of acoustic waves is suppressed, as it is in a Boussinesq fluid, in which the background density variations are also neglected.) Although the details of these calculations did not match the observed properties of the solar cycle in several important respects (see below), they demonstrated that the dynamo process may work in this way in other stars.

A crucial property of the models is that the form of the motion is determined by the Coriolis force, which leads to convective cells elongated parallel to the axis of rotation. Convection experiments carried out in space (so as to obtain zero gravity environment) confirm that so-called *banana* cells appear when the rotation rate is sufficiently large (Hart *et al.* 1986). In such a configuration, the angular velocity tends to be constant on approximately cylindrical surfaces (§11.4). Jones, Galloway, and Roberts (1990) have, on the other hand, shown that, in the presence of a toroidal magnetic field B, the mode of convection is determined by the Elsasser number Λ, given by

$$\Lambda = \frac{B^2}{2\mu\rho\eta\Omega}.$$

For small values of Λ banana cells are favoured, but for values in excess of a critical value (~ 1), the onset of convection should take the form of azimuthal rolls centered about the rotation axis.

Brandenburg *et al.* (1992) and Nordlund *et al.* (1992) have attempted to model turbulent dynamos with some striking results. They simulate compressible convection in a stratified, rotating system and show that the magnetic field can be maintained by dynamo action, if the magnetic Reynolds number is sufficiently large. The resulting field is highly intermittent: the flux tubes wind round isolated sinking plumes with locally concentrated vorticity. There is no systematic cyclic behaviour, but these results show that turbulent convection in a rotating star is likely to generate small-scale magnetic fields.

11.7 Fast dynamo models

Although numerical solutions of the non-linear dynamo equations provide valuable insights into the ways in which astrophysical dynamos *might* operate, they can never represent the physical conditions actually existing in the Sun, since the solar plasma is too inviscid.

It was shown that the possibility of dynamo action involves a delicate interaction between the source term in Equation (11.2), $\nabla \times (\mathbf{u} \times \mathbf{B})$, and the ohmic decay or diffusivity term, $\eta \nabla^2 \mathbf{B}$. While the source term must be of sufficient magnitude to overcome the effects of ohmic decay, as expressed in the diffusivity term, the dynamo cannot operate in a region of zero diffusivity, for it then remains *frozen-in* to the plasma. Though the field may grow and stretch, it cannot move relative to the plasma.

To achieve a dynamo, the enhanced field must, in some way, be released from the element of plasma involved in the stretching and transferred to another element. For the Sun, unfortunately, the molecular diffusivity is too small to permit the fields to 'slip away' and the simulations described in § 11.5 indicate that the solar magnetic fields are too strong to allow the concept of decay via the mechanism of turbulent diffusion to be valid. The concept of the *fast dynamo* has been introduced in order to find flows for which the dynamo retains a finite growth rate in the limit $\eta \to 0$.

Consider again the basic definition of dynamo action given in §11.1, i.e. if $\mathbf{B} = \mathbf{B}_0(\mathbf{r}) \exp(qt)$, $\mathrm{Re}(q) \geq 0$. All kinematic dynamos which have been investigated so far have the property that $\mathrm{Re}(q)t_0 \to 0$ as $R_m \to \infty$ (or $\eta \to 0$); i.e. they die out (and are said to be *slow*), or they have artificial singularities in velocity or vorticity responsible for their maintenance. A *fast dynamo*, by contrast, has the property that $\mathrm{Re}(q)t_0 \to$ constant $\neq 0$ as $R_m \to \infty$, and, for such dynamos, diffusion is irrelevant.

The challenge of the fast dynamo is therefore to compute the growth rate $\mathrm{Re}(q)$ as a function of R_m and show that it approaches a constant value as R_m increases without limit. The classic example of such a dynamo is the 'stretch-twist-fold' dynamo, first studied by Vainshtein and Zel'dovich (1972), which operates independently of the diffusivity but is of little relevance to astrophysical plasmas. Galloway and Proctor (1992) have investigated flows which are chaotic (in the mathematical sense) and found examples in which the magnetic field evolution strongly suggests that the dynamos are fast. While the flows are again unlikely to be of relevance to astrophysical situations, they demonstrate that fast dynamos exist for non-singular flows with realistic values of η. The question whether such dynamos can exist in situations in which the Lorentz force must be included is, however, entirely unanswered.

11.8 The small-scale eddies

Another approach considers the back-reaction on the small-scale fluid flows, which cannot be truly modelled by even the most sophisticated numerical simulations. Vainshtein, Parker, and Rosner (1992) show that this back-reaction can completely modify the nature of the field amplification. They point out that, in three dimensions, the turbulent diffusion cannot be completely suppressed, since a three-dimensional fluid has the freedom to arrange itself so that the magnetic back-reactions are minimized.

The fluid motions can, in this way, become two-dimensional in planes locally orthogonal to the strong magnetic field lines. Such motions, which correspond to small-scale interchange of field lines, are not inhibited by the Lorentz force. It will be of interest to look for the 'two-dimensionalization' of turbulent flows at the small spatial scales at which magnetic fields first become dynamically important and to study the effects on turbulent diffusion of magnetic fields resulting from this behaviour.

11.9 Conclusion

There have been many other studies of linear and non-linear dynamos which cannot be discussed in detail here. There can be no objection, in principle, to invoking dynamo action in astrophysical situations, but the details of the dynamo operation are still unclear in many cases.

It would seem that, under circumstances where the assumptions of mean-field electrodynamics apply, dynamo action in astrophysical situations may be invoked with confidence. These assumptions may well apply to stars with shallow convection zones, such as those described by Gray (1992), whose argument that an α^2 dynamo is operating in these stars is compelling.

The assumptions do not, however, apply to solar magnetic fields, where the perturbation field strength exceeds that of the mean field by several orders of magnitude, and where the essential part which must be played by turbulent diffusivity can only be shown to work for fields considerably weaker than the large-scale fields of the Sun. It would seem likely that similar reservations must apply to the magnetic fields of many solar-type stars.

On the other hand, since the α–ω dynamo depends for its existence on cyclonic convection to provide the α-effect and non-uniform rotation for the ω-effect, the dependence of cyclic and non-cyclic behaviour in stars on an appropriate balance between depth of the convection zone and rotation period (see §7.8) is entirely consistent with the involvement of dynamo action in one or more of its various forms.

In the next chapter we review some recent results from helioseismology which have raised additional difficulties in understanding the details of how a dynamo might operate in the Sun.

References

Brandenburg, A., Jennings, R. L, Nordlund, Å., and Stern, R. F.: 1992, in *Spontaneous Formation of Space-Time Structures and Criticality*, eds. T. Riste and D. Sherrington, Kluwer, Dordrecht, (in press).

Cattaneo, F., and Vainshtein, S. I.: 1991, *Astrophys. J.*, **376**, L21.

Cowling, T. G.: 1976, *Magnetohydrodynamics*, Adam Hilger, Bristol.

Cowling, T. G.: 1981, *Ann. Rev. Astron. Astrophys.*, **19**, 115–35.

Galloway, D. J., and Proctor, M. R. E: 1992, *Nature*, **356**, 691.

Gilman, P. A.: 1983, *Astrophys. J. Supp.*, **53**, 243.

Gilman, P. A., and Miller, J.: 1986, *Astrophys. J. Supp.*, **61**, 585.

Glatzmaier, G. A.: 1985, *Astrophys. J.*, **291**, 300.

Glatzmaier, G. A.: 1985, *Geophys. Astrophys. Fluid Dynam.*, **31**, 137.

Golub, L., Rosner, R., Vaiana, G. S., and Weiss, N. O.: 1981, *Astrophys. J.*, **243**, 309.

Gray, D. F.:1992, in *Proceedings of the National Solar Observatory/Sacramento Peak 12th Summer Workshop*, ed. K. L. Harvey, San Francisco, California, 543.

Hart, J. E., Toomre, J., Deane, A. E., Hurlburt, N. E., Glatzmaier, G. A., Fichtl, G. H., Leslie, F., Fowlis, W. W., and Gilman, P. A.: 1986, *Science*, **234**, 61–4.

Hart, J. E., Glatzmaier, G. A., and Toomre, J.: 1986, *J. Fl. Mech*, **173**, 519–44.

Jennings, R. L.: 1992, in *Proceedings of the National Solar Observatory/Sacramen-to Peak 12th Summer Workshop*, ed. K. L. Harvey, San Francisco, California, 543.

Jones, C. A., Weiss, N. O., and Cattaneo, F.: 1985, *Physica*, **14D**, 161.

Jones, C. A., Galloway, D. J., and Roberts, P. H.: 1990, *Geophys. Astrophys. Fluid Dynam.*, **53**, 145–182.

Kleeorin, N. L., and Ruzmaikin, A. A.: 1991, *Solar Phys.*, **131**, 211.

Krause, F., and Rädler, K. H.: 1980, *Mean-field Magnetohydrodynamics and Dynamo Theory*, Pergamon, Oxford.

Leighton, R. B.: 1969, *Astrophys. J.*, **156**, 1.

Moffatt, H. K.: 1974, *J. Fluid Mech.*, **65**, 1–10.

Moffatt, H. K.: 1976, *Adv. Applied Mech.*, **16**, 119–81.

Moss, D., Tuominen, I., and Brandenburg, A.: 1990, *Astron. and Astrophys.*, **228**, 284.

Nordlund, A., Brandenburg, A., Jennings, R. L., Rieutord, M., Ruokolainen, J., Stein, R. F., and Tuominen, I.: 1992, *Astrophys. J.*, (in press).

Noyes, R. W., Hartmann, L. W., Baliunas, S. L., Duncan, D. K., and Vaughan, A. H.: 1984, *Astrophys. J.*, **279**, 763.

Parker, E. N.: 1955, *Astrophys. J.*, **122**, 293.

Parker, E. N.: 1970, *Astrophys. J.*, **160**, 701–24.

Parker, E. N.: 1975, *Astrophys. J.*, **198**, 205.

Parker, E. N.: 1979, *Cosmical Magnetic Fields: Their Origin and Their Activity*, Clarendon Press, Oxford.

Parker, E. N.: 1987a, *Solar Phys.*, **110**, 11.

Parker, E. N.: 1987b, *Astrophys. J.*, **312**, 868.

Piddington, J. H.: 1972, *Solar Phys.*, **22**, 3.

Roberts, G. O.: 1971, *Phil. Trans. A.*, **271**, 412–454.

Rosner, R., and Weiss, N. O.: 1992, in *Proceedings of the National Solar Observatory/Sacramento Peak 12th Summer Workshop*, ed. K. L. Harvey, San Francisco, California, 511.

Schmitt, D.: 1987, *Astron. and Astrophys.*, **174**, 281.

Stix, M.: 1987, *Solar & Stellar physics*, ed. E-H. Schroter and M. Schüssler, Springer, Berlin, 15.

Stix, M.: 1989, *The Sun: an Introduction*, Springer, Berlin.

Vainshtein, S. I., Parker, E. N., and Rosner, R.: 1992, *Astrophys. J.*, submitted.

Vainshtein, S. I., and Zel'dovich, Ya. B.: 1972, *Soviet Phys. Uspekhi*, **15**, 159.

Weiss, N. O., Cattaneo, F. and Jones, C. A.: 1984, *Geophys. & Astrophys. Fluid Dyn.*, **30**, 305–41.

Yoshimura, H.: 1976, *Solar Phys.*, **50**, 3–23.

Yoshimura, H.: 1981, *Astrophys. J.*, **247**, 1102–12.

12

Helioseismology and the solar cycle

12.1 Introduction

The measurement and interpretation of the travel times of earthquake signals have been used for many years to study the structure of the Earth's interior. In addition to the classical method, observations of free oscillations have been used since the great Chilean earthquake of 1960, and this branch of seismology has been recently applied to the Sun with considerable success. For a given solar model, the eigenfunctions and eigenfrequencies of the radial velocity perturbations may be calculated by standard methods and compared to the observed frequencies. Improved models of the internal structure of the Sun may be obtained by bringing as many calculated frequencies as possible into agreement with observed frequencies. Such calculations have, for example, forced the rejection of 'low-Z' models of the solar interior which, it had been hoped, might resolve the solar neutrino problem (Christensen-Dalsgaard and Gough 1982).

Of particular interest for studies of cyclic phenomena is the rotational modulation of the eigenmode frequencies. Because the eigenfunctions of the many non-radial p-modes exhibit different depth and latitude dependence, the measurement of frequency splittings in the intermediate orders of the waves has led to inferences regarding the internal rotation rate as a function of depth and latitude (Duvall *et al.* 1986, Brown and Morrow 1987, Morrow 1988, Brown *et al.* 1989, Libbrecht 1989). Surprisingly, these results at first suggested that, within certain error bars, the angular velocity is independent of radial distance across the convection zone; i.e. that the differential rotation (with latitude) at the base of the convection zone is very similar to that at the surface. Brown *et al.* (1989) have concluded that 'this confronts both α–ω dynamo models and current dynamical theories of the convection zone with severe difficulties'.

In this chapter the nature of these difficulties is explored, and the above and other models of the internal rotation of the Sun are investigated in the light of the latest p-mode splitting data. Cyclic variations of the parameters measured by helioseismologists are also discussed.

12.2 The iso-rotation surfaces

An *iso-rotation surface* is defined to be a geometrical surface on which the angular velocity of rotation is constant. On the assumption that these surfaces are axially symmetric, they may be represented by two-dimensional contours, such as those shown in Figure 12.1, but it should be remembered that significant departures from axisymmetry may occur. Figure 12.1 shows a selection of six rotation models in which the iso-rotation surfaces are represented by contour lines on a section of the Sun through the polar axis. The grey-scale indicates regions of slower-than-average rotation by dark shading and faster-than-average by light shading. These models will be discussed during this chapter.

In the previous chapter, it was shown that, if the helicity is negative (i.e. α is positive), a migratory dynamo wave will propagate along an iso-rotation surface in a direction which forms a right-handed set with the normal to the surface and the direction of rotation; if the helicity is positive, they form a left-handed set (see § 11.4).

The non-linear dynamical calculations of Gilman (1983), Gilman and Miller (1986), Glatzmaier (1985), and others indicate that the angular velocity of rotation should increase outwards through the convection zone, being constant on cylinders of constant radius about the rotation axis. The corresponding iso-rotation surfaces are similar to those illustrated as Model 1 in Figure 12.1 and, if one assumes negative helicity, the right-hand rule indicates that a dynamo located within the convection zone should propagate polewards, a result which is contrary to the evidence of the butterfly diagram.

Snodgrass and Wilson (1987) have argued that the torsional oscillation signal is the surface manifestation of the Coriolis effect on a system of azimuthal or 'doughnut-shaped' rolls, symmetric about the rotation axis, which organize the convective flows at high latitudes, and that the maximum shear zone of the torsional oscillation pattern corresponds to the highly localized (in latitude) downflow of the rolls. Conservation of angular momentum within these downflows thus gives rise to an inward increase in the rotational velocity, as in Model 2 (see Figure 12.1). Application of the right-hand rule then results in a dynamo wave propagating away from the pole along these surfaces and emerging, appropriately, in the sunspot zone.

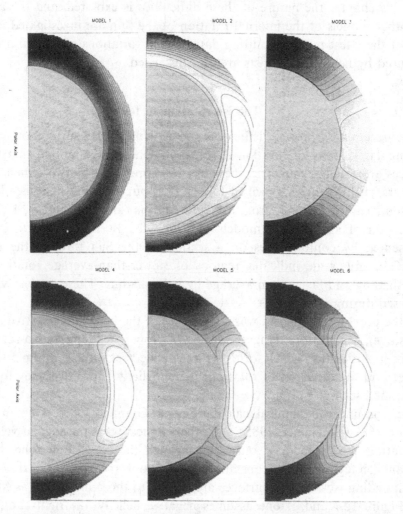

Fig. 12.1 The iso-rotation surfaces corresponding to Models 1–6 (as defined in the text) are represented in meridional section by the contour lines drawn within the region $0.7 < r/R < 1$. The grey-scale indicates regions of slower-than-average rotation by dark shading and faster-than-average by light shading. The vertical axis corresponds to the rotation axis.

It is clear that independent information regarding the iso-rotation surfaces of the solar interior is crucial to the dynamo problem, and the frequency splittings of the non-radial p-modes offer an exciting prospect.

12.3 The oscillations and the observed frequency-splittings

Solar oscillations have been observed either as white-light fluctuations or as Doppler shifts arising from acoustic-mode oscillations in a particular spectral line. An individual acoustic mode is defined in terms of the displacement vector $\delta\mathbf{r}_{n\ell m}(r, \theta, \phi, t)$ by

$$\delta\mathbf{r}_{n\ell m} = \mathrm{Re}\left\{\left[\xi_{n\ell}(r)P_\ell^m(\cos\theta)\mathbf{a_r} + L^{-1}\eta_{n\ell}(r)\right.\right.$$
$$\left.\left.\left(\frac{dP_\ell^m}{d\theta}\mathbf{a}_\theta + \frac{im}{\sin\theta}P_\ell^m(\cos\theta)\mathbf{a}_\phi\right)\right]\ \exp\{i(m\phi - 2\pi\nu_{n\ell m}t)\}\right\} \quad (12.1)$$

(Brown *et al.* 1989). Here $\mathbf{a_r}, \mathbf{a}_\theta$, and $\mathbf{a}_\phi$ are unit vectors in the r, θ, and ϕ directions; t is time; $L^2 = \ell(\ell + 1)$; and $\xi_{n\ell}(r)$ and $\eta_{n\ell}(r)$ are the amplitudes of the radial and horizontal displacements, which may be obtained by solving the adiabatic oscillation equations. Each oscillation mode is therefore characterized by the eigenvalues n, ℓ, and m, and the associated frequency $\nu_{n\ell m}$. The solar oscillations appear in $(n\ell)$ multiplets and, in the absence of rotation and magnetic field, would exhibit a $(2\ell + 1)$-fold degeneracy in m. In a slowly rotating star, such as the Sun, rotation removes the degeneracy, causing a nearly uniformly spaced fine structure in each multiplet. The spacing arises from the Doppler effect of the rotation on the wave frequencies and may therefore be used as a measure of the local angular velocity. Helioseismology relies on fine-structure data from many such multiplets to provide information from different samplings of the solar interior.

In general, stellar oscillations are non-radial and non-adiabatic, and their properties are calculated by perturbing and solving the equations of continuity, hydrostatic equilibrium, conservation of energy, and radiative flux transport (Unno *et al.* 1989). The solar oscillation modes which have been clearly identified have periods of about five minutes and range in degree from $\ell \sim 10$ to $\ell \sim 100$. Their properties are described by the asymptotic equation for the amplitude f (Goode 1992),

$$f'' + f\left\{\left(\frac{2\pi\nu r}{c}\right)^2 - \ell(\ell + 1) + O\left(\left|\frac{d\ln\rho}{d\ln r}\right|^2, 1\right)\right\} = 0, \quad (12.2)$$

where ρ is density and c is the sound speed. The inner reflection point for these sound waves occurs when the first two terms in the brackets are comparable, such that, the lower the degree of the oscillation, the more deeply it samples the physical or rotational conditions of the interior. The lowest-degree modes ($\ell < 5$) sample all the way to the Sun's centre. Near the surface, the reflection point for these modes occurs where the third term in the brackets is comparable with the first, so that the interaction of the oscillations with the surface is essentially independent of ℓ.

In the earlier data obtained by Duvall *et al.* (1986) and by Brown and Morrow (1987), a cross-correlation technique was used which effectively averaged the frequency shifts over n to yield $v'_{\ell m} = \bar{v}_{\ell m} - \bar{v}_{\ell 0}$. For each ℓ there are $2\ell + 1$ values of $v'_{\ell m}$ but, in view of probable inaccuracies in the data, it is not profitable to analyse these on an individual basis (particularly when ℓ is large), and they are generally fitted by least squares to a series of Legendre polynomials

$$v'_{\ell m} = \bar{v}_{\ell m} - \bar{v}_{\ell 0} = \ell \sum_{k=0}^{5} a_k^\ell P_k\left(\frac{m}{\ell}\right),\tag{12.3}$$

where only the odd coefficients a_1^ℓ, a_3^ℓ, and a_5^ℓ provide information about the rotational splittings. The even-order coefficients are not directly associated with any perturbation beyond the second-order effects of rotation, such as distortion. It is generally assumed, however, that part of the symmetric signal arises from the effects of magnetic fields in active latitudes, where the perturbations would have only a small ℓ-dependence (see §12.10). Observed values of the odd coefficients obtained by Brown and Morrow (1987) are reproduced in Figure 12.2.

The best data available to date are from the Big Bear Solar Observatory (BBSO) (Libbrecht 1989, Libbrecht and Woodard 1990), where the data sets are 100 days long and the signal-to-noise ratio is such that it is not necessary to average over n. By a similar Legendre polynomial expansion, Libbrecht obtained the splitting coefficients $a_1^{n\ell}$, $a_3^{n\ell}$, and $a_5^{n\ell}$. Instead of averaging over n, however, Libbrecht noted that the scatter in the $a_k^{n\ell}$ fits depended strongly on the frequency $v_{n\ell}$ but was least in the neighbourhood of 2.5 mHz. He therefore fitted the $a_k^{n\ell}$ by least squares to the relation

$$a_k^{n\ell} = a_k^*(\ell) + b_k^*(\ell)(v_{n\ell} - 2.5),\tag{12.4}$$

and his results for the odd coefficients $a_k^*(\ell)$, $k = 1, 3, 5$, are shown in Figure 12.3.

12.4 The interior rotation profile and the theoretical splittings

It is convenient to represent the Sun's angular velocity $\Omega(r, \theta)$ by the expansion

$$\Omega(r, \theta) = \sum_{s=0}^{2} \Omega_s(r)\mu^{2s},\tag{12.5}$$

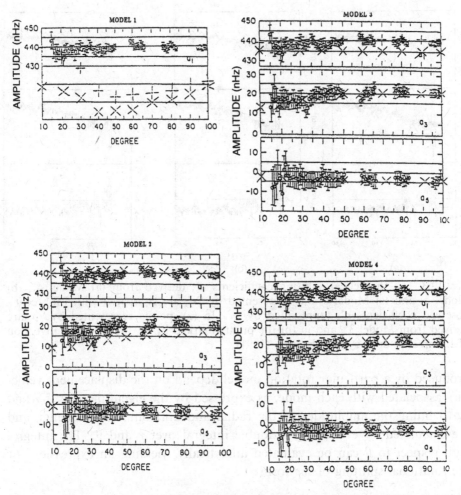

Fig. 12.2 The theoretical splitting coefficients a_k^*, derived from Models 1–4, and represented by '×' or '+', are compared with the observed coefficients a_k^j of Brown and Morrow (1987), represented by 'Φ'. For Model 1, points marked '+' represents the case in which ω_i is increased to 480. For Model 3, points marked '+' represent the case in which ω_0 is increased to 458.

where $\mu = \cos\theta$. (Alternatively, it may be represented by a truncated series of orthogonal Legendre polynomials

$$\Omega(r,\theta) = \sum_{s=0}^{2} \Omega_s^*(r) P_{2s}(\cos\theta),$$

and it is easy to establish the relations between the Ω_s and Ω_s^* which provide for the equivalence of these two representations.) The frequency splittings

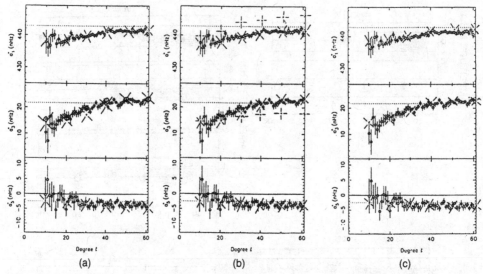

Fig. 12.3 The theoretical splitting coefficients a_k^*, derived from (a) Model 4*, (b) Model 5, and (c) Model 6 and represented by '×', are compared with Libbrecht's observed coefficients a_k^*, represented by 'ϕ'. In (a), points represented by '$\odot$' are derived from Model 3*, and in (b) the points marked by '+' are from the inversion of Libbrecht's data by Goode *et al.* (1991).

produced by a particular rotation model depend on the displacement velocities associated with each mode, as expressed by the eigenfunctions $\xi_{n\ell}$ and $\eta_{n\ell}$ defining the amplitudes of the radial and horizontal components, and are calculated by evaluating a double integral over r and θ. The integral with respect to θ can be evaluated analytically, yielding an expression for the rotational splitting of the form

$$v'_{n\ell m} = \frac{m}{2\pi} \sum_{s=0}^{2} \int_0^R K_{n\ell ms}\Omega_s(r)\mathrm{d}r, \qquad (12.6)$$

where R is the solar radius and $K_{n\ell ms}$ is a rotation kernel depending on the eigenfunctions of the oscillation modes and on the solar model. Expanding in the odd Legendre polynomials yields

$$v'_{n\ell m} = \ell \sum_{k=0}^{2} a_{2k+1}^{n\ell} P_{2k+1}\left(\frac{m}{\ell}\right), \qquad (12.7)$$

and, in the limit of large ℓ, Brown *et al.* (1989) have derived the expressions

$$a_1^{n\ell} \approx \frac{1}{2\pi} \int_0^R \left[\Omega_0(r) + \frac{1}{5}\Omega_1(r) + \frac{3}{35}\Omega_2(r)\right] H_{n\ell}(r)\mathrm{d}r, \qquad (12.8)$$

$$a_3^{n\ell} \approx -\frac{1}{10\pi} \int_0^R \left[\Omega_1(r) + \frac{2}{3}\Omega_2(r) \right] H_{n\ell}(r)\mathrm{d}r, \tag{12.9}$$

$$a_5^{n\ell} \approx \frac{1}{42\pi} \int_0^R \Omega_2(r)H_{n\ell}(r)\mathrm{d}r, \tag{12.10}$$

where

$$H_{n\ell}(r) = \frac{\{\xi_{n\ell}^2(r) + \eta_{n\ell}^2(r)\}\rho(r)r^2}{\int_0^R \{\xi_{n\ell}^2(r) + \eta_{n\ell}^2(r)\}\rho(r)r^2\mathrm{d}r}. \tag{12.11}$$

Linear combinations of Equations (12.8), (12.9) and (12.10) yield

$$a_1^{n\ell} + a_3^{n\ell} + a_5^{n\ell} \approx \frac{1}{2\pi} \int_0^R \Omega_0(r)H_{n\ell}(r)\mathrm{d}r = \langle \Omega_0 \rangle = W_0, \tag{12.12}$$

$$-5a_3^{n\ell} - 14a_5^{n\ell} \approx \frac{1}{2\pi} \int_0^R \Omega_1(r)H_{n\ell}(r)\mathrm{d}r = \langle \Omega_1 \rangle = W_1, \tag{12.13}$$

and

$$21a_5^{n\ell} \approx \frac{1}{2\pi} \int_0^R \Omega_2(r)H_{n\ell}(r)\mathrm{d}r = \langle \Omega_2 \rangle = W_2, \tag{12.14}$$

where the angular brackets denote an average over r, weighted with $H_{n\ell}$.

Since the splitting coefficients a_k^ℓ of Brown and Morrow (1987) were effectively averaged over n, they have rewritten the above integral equations in terms of an averaged kernel $H(r; w)$ in place of $H_{n\ell}$. Here

$$H(r; w) = \frac{(1 - w^{-2}a^2)^{-\frac{1}{2}}c^{-1}}{\int_{r_t}^R (1 - w^{-2}a^2)^{-\frac{1}{2}}c^{-1}\mathrm{d}r}, \tag{12.15}$$

where w is the reduced frequency, $2\pi\nu_{n\ell 0}/\sqrt{\ell(\ell+1)}$, and $a = c/r$, where c is the sound speed. The lower limit of integration is given by $a(r_t) = w$.

Libbrecht's data are of sufficient accuracy to determine the individual frequency splittings for each set of n, ℓ, m and hence to determine each $a_k^{n\ell}$, but here the problem is an apparent excess of data. Not only is the task of solving these equations for every $n\ell$ pair extremely tedious, but it may introduce needless noise into the solutions, since Equation (12.2) shows that it is the different values of ℓ which permit the sampling of different regions of the solar interior. For this reason, Libbrecht expressed his results in terms of the $a_k^*(\ell)$-coefficients (Equation 12.4), and it is necessary to evaluate the integrals only for the values of n for which $\nu_{n\ell} \approx 2.5$ mHz in order to determine theoretical values of a_k^* for different rotation models.

The aim of the analysis is to find the Ω_k as functions of r, and, to this end, Equations (12.12), (12.13), and (12.14) must be solved for $\Omega_1(r)$, $\Omega_3(r)$, and $\Omega_5(r)$, either by inverting the three integrals (the *inverse* approach) or by postulating analytical or numerical models for $\Omega_1(r)$, $\Omega_3(r)$, and $\Omega_5(r)$ and fine-tuning them until the computed values of the $a_k^{n\ell}$ fall within the error bars of the observed values (the *forward* approach).

These methods have been extensively discussed by Morrow (1988), Brown *et al.* (1989), Goode *et al.* (1991), and others. The inverse problem, that of deriving the rotation profile function, $\Omega(r, \theta)$, directly from the frequency data, has been attempted by many writers. Gough (1984) has pointed out that the kernel $H(r; w)$, defined by Equation (12.15), can easily be transformed into an Abel kernel, and that, consequently, if data spanning a sufficient range of w were available, Equations (12.12)–(12.14), with $H_{n\ell}$ replaced by $H(r; w)$ could be inverted to yield Ω in terms of the observed coefficients.

To perform a complete formal inversion of the equations would require modes with w as low as $a_s = c(R)/R$, which are not available. An inversion in the range of the turning points r_t of the available data can, however, be carried out in terms of the angular velocity Ω at radii greater than the greatest value r_m of r_t. With the lowest value of w for which data are available denoted by $w_m = a(r_m)$, the result (Brown *et al.* 1989) is

$$-\frac{\pi}{2}\Omega_j(r) = a^3 \frac{\mathrm{d}}{\mathrm{d}\ln r} \int_{r_t}^{R} \frac{G_j(w)\mathrm{d}w}{w^2(a^2 - w^2)^{\frac{1}{2}}} + \Psi_j(r; w_m), \qquad (12.16)$$

where

$$G_j(w) = W_j(w) \int_{r_t}^{R} (1 - w^{-2}a^2)^{-\frac{1}{2}}c^{-1}\mathrm{d}r, \qquad (12.17)$$

and the W_j are defined in Equations (12.12)–(12.14). The function $\Psi(r; w)$ is a correction term that takes account of the contribution from the rotation for $r_m \le r \le R$.

In principle, the quantity G_j is available only at discrete values of w. In practice, even that is not available, since the observed splitting coefficients a_k^ℓ are averages over n and adjacent ℓ, and hence over w. So, in order to estimate $\Omega_k(r)$ from Equations (12.16) and (12.17), $W_k(w)$ must be represented by smoothing functions computed from the averages of the discrete values. The inversion methods are, therefore, not without problems. Further, noise is inevitably present in the data, and this can cause instability, in the sense that similar realizations of the frequency splittings may be mapped into widely differing models of the rotational velocity, a process which entails the risk of spurious or non-physical angular velocities (Morrow 1988).

In the forward approach, the theoretical frequency-splitting coefficients are computed for various models of the rotation profile and are compared with the observed coefficients. Although not without its own share of problems (e.g. the question of uniqueness), this approach permits the study of a wide variety of rotation models which may be suggested by other considerations, such as the need to study rotation models compatible with the α–ω dynamo. If a model maps well into the data, it may be considered as one of (possibly) several candidates for representing a physical situation which is consistent with the data. Depending on other physical considerations, one might invest such a model with a greater or lesser degree of confidence, but one cannot, of course, claim uniqueness for it.

Perhaps the greatest strength of the forward method is that, when an otherwise plausible model does not map to within the error bars of the data, then, provided one is able to trust the observer's error bars, that particular model may be eliminated.

12.5 Results

Brown *et al.* (1989) have applied both the forward and the inverse methods to the analysis of the data of Brown and Morrow (1987) and demonstrated convincingly that models for which the angular velocity increases outwards through the convection zone, being constant on cylinders of constant radius about the rotation axis, are untenable (see also § 12.6 below).

They conclude, in fact, that the latitudinal angular velocity observed at the surface varies little through the convection zone. This result would imply iso-rotation surfaces of the form illustrated by Model 3 in Figure 12.1, the normal to the surfaces being in the meridional direction and in the equatorward sense throughout the convection zone. They also deduce that, in the equatorial region, Ω decreases with radius for $r < r_m \approx 0.89$, but, nearer the rotation axis, Ω increases. In the vicinity of $r = 0.5$, the polar and equatorial rates are about the same, but Brown and Morrow's data are not sufficiently reliable at low ℓ to permit inferences at lower values of r. Libbrecht (1989), however, has reported that his data from the Big Bear Observatory support the conclusion that the rotation rate varies only slightly with radius across the convection zone.

Goode *et al.* (1991) have inverted all the available oscillations data by the same method and found that the rotation rate in the equatorial plane declines going inwards from the surface to $r = 0.6$, and, with less certainty, that the polar rate increases to the same depth, so that the surface differential rotation decreases with depth. The bulk of these changes, they deduce, occur near the base of the convection zone. On the assumption that the Sun should rotate rigidly below the convection zone, Gilman and others have also inferred the existence of a shallow layer or 'ramp' (Morrow 1988) between the convection zone and the rigidly rotating interior.

As a result of these and other analyses, it is now widely believed (e.g. Rosner and Weiss 1992) that the surface differential rotation profile extends uniformly to the bottom of the convection zone, below which exists a shallow transition or overshoot layer. If this rotation model is correct, application of the right-hand rule (see § 11.4) would require that any dynamo wave generated within the convection zone should propagate radially outwards, a result inconsistent with the patterns of activity exhibited by the Sun.

In the transition zone, however, the angular velocity gradients may be large. The data indicate an interior rotation period of $\sim$ 27 days, corresponding to the surface rate at latitudes $\pm 37°$, such that, across the transition zone, the radial gradient should be positive in the latitude range $0° - 37°$ and negative between latitude $37°$ and the poles, as is the case with Model 3 (see Figure 12.1). Gilman *et al.* (1989) argued that, within this zone, the helicity should be positive (i.e. α negative) and that the dynamo, if it exists, must operate there. The left-hand rule (see § 11.4) then yields two waves, a lower latitude wave propagating equatorwards and a higher latitude wave propagating polewards. How such strong gradients can be produced within this transition or overshoot zone, while they are apparently absent within the convecting region above, where rotation must generate a strong helicity signal, is a problem which theoreticians have been unable to understand.

Nevertheless, this model is currently the most popular choice, and other variants of the MFE kinematic α–ω dynamo have been proposed based on similar rotation models. Typically, strong toroidal fields accumulate in the region of convective overshoot and support magnetostrophic waves which are excited by instabilities, powered by magnetic buoyancy. These waves have a net helicity which provides an α-effect that is antisymmetric about the equator and concentrated near the base of the convection zone; α is negative at low-latitudes in the northern hemisphere but changes sign at higher latitudes, where it drops in magnitude. This combination permits the production of dynamo waves which migrate towards the equator. The introduction of a radial dependence of the turbulent diffusivity η, such that it falls to zero at the base of the overshoot layer, ensures that the strong fields are concentrated there.

12.6 Results of forward calculations

To investigate the possibility that there may be other models which are also compatible with the frequency splitting data, the author (Wilson 1992) has explored several models and compared the frequency splittings calculated from a forward analysis with the data of Brown and Morrow (1987).

Using Equation (12.5) it is possible to generate many different models of the iso-rotation surfaces within the Sun, a selection of which is shown in Figure 12.1. In each of these models, Ω_1 and Ω_2 are set to zero below $r/R = 0.7$, while Ω_0 is set to a constant, ω_i, which may be chosen arbitrarily. Above this level, the models are defined by straightforward analytic functions, subject to the surface boundary condition of the form

$$\frac{\Omega(R, \theta)}{2\pi} = \omega_0 - \omega_1 \mu^2 - \omega_2 \mu^4. \tag{12.18}$$

For the observed plasma rotation rate (Snodgrass, 1984), $\omega_0 = 452$, $\omega_1 = 49$, and $\omega_2 = 84$ (in nHz).

On the basis of these models, theoretical values of $a_k^*(\ell)$ may be calculated from Equations (12.8)–(12.10), using values of n for which $\nu_{n\ell} = 2.5$ mHz. These are compared with the values of a_k^ℓ obtained by Brown and Morrow in Figure 12.2. Model 1, in which the angular velocity is (approximately) constant on concentric cylinders, is obviously inconsistent with the observational data, the values of $a_1^*(\ell)$ being consistently too low. This result reflects an essential feature of the model: that, if the rotation rate is increasing outward, it must be less than the surface rate through most of the convection zone. If the interior rotation rate ω_i is regarded as a free parameter and increased arbitrarily, the low ℓ-values of $a_1^*(\ell)$ may be raised to be compatible with a_1^ℓ, but they fall off rapidly at higher ℓ-values. Values of $a_3^*(\ell)$ and $a_5^*(\ell)$ are also incompatible with the corresponding values of a_j^ℓ but are not shown.

Model 2, which is consistent with the azimuthal roll model of Snodgrass and Wilson (1987), fares somewhat better and, by minor parameter variation, it is relatively easy to match the values of a_1^ℓ. Values of $a_3^*(\ell)$ and $a_5^*(\ell)$ are, however, consistently less than the corresponding a_k^ℓ, a consequence of the steady increase of $\Omega(r, 0)$ with depth at higher latitudes.

Using Model 3, it is possible to match the values of a_3^ℓ and a_5^ℓ satisfactorily, but it is a little surprising to find that this model yields values of $a_1^*(\ell)$ which lie consistently below the error bars of the observed values of a_1^ℓ. Brown *et al.* (1989, p. 534) have obtained their optimal results for this type of model by arbitrarily increasing ω_0, the surface equatorial rate, and reasonable agreement for Model 3 may be obtained with $\omega_0 = 458$ (cf. their value of 457), a value intermediate between Snodgrass's tracer rate (462) and his plasma rate, which is also indicated in Figure 12.2 by the points marked '+'.

However, the equatorial angular velocity should not be regarded as a free parameter in the analysis of frequency splitting data (as can the interior rate), since it may be derived independently from surface spectroscopic data. It therefore must be concluded that either the surface plasma rotation rate given by Snodgrass is too low, or that models in which the surface differential rotation profile extends to the base of the convection zone are *not* compatible with current data. Since the values of $a_1^*(\ell)$ lie consistently below those for a_1^ℓ, the velocity profile *must* increase initially with depth, at least at lower latitudes. This conclusion may also be inferred by comparing spectroscopic rotation rates with those derived from tracers, such as sunspots and, in particular, the supergranule patterns. These latter yield the fastest rotation rates at all latitudes, and it is assumed that they rotate at rates associated with the depths at which they are formed (or anchored).

These considerations led to the development of Model 4, for which the angular velocity increases initially with depth up to latitudes 45°. For this model, the theoretical values of the $a_k^*(\ell)$ lie comfortably within the error bars of the Brown and Morrow values of a_j^ℓ, and this is typical of a range of models of this type.

It is thus incorrect to conclude that results from helioseismology imply that the differential rotation profile is necessarily independent of depth throughout the convection zone. It is sufficient to note that the early results were not inconsistent with such a model.

12.7 Comparison with Libbrecht's data

The error bars on the data points of Brown and Morrow are rather generous, giving scope for the incorporation of a variety of models within them. Libbrecht's (1989) data (Figure 12.3) claim considerably higher precision at values of $\ell >\sim 30$, and these place more severe constraints on the possible models.

A study of these data points, however, shows significant variations between adjacent ℓ-values, e.g. $a_1^*(21) - a_3^*(23)$ and $a_3^*(32) - a_3^*(34)$, raising questions as to whether these variations should be interpreted as being of solar origin or as indicating some additional uncertainties, or noise, within the data. The contribution functions for selected ℓ-values are obtained by evaluating $H_{nl}(r)$ (Equation 12.11) and show considerable overlap (see Figure 12.4); a solar interpretation of these variations would therefore require that quite significant variations in the physical structure of the convection zone be maintained over a narrow depth range for periods of ~ 100 days (the length of Libbrecht's data run), a pattern which seems unlikely. The error limits claimed by Libbrecht should therefore be regarded with some caution.

If the spread of the data points is taken to define the error bars, the constraints are relaxed but are, nevertheless, more severe than those provided by the data of Brown and Morrow. The author has tested further models in an attempt to fit Libbrecht's data within these limits. Results for modified versions of Models 3 and 4, Models 3* and 4*, and for two further models, Models 5 and 6, are compared with Libbrecht's points in Figure 12.3.

Points calculated from Model 6 lie either within Libbrecht's error bars or within the spread of adjacent data points for all values of ℓ computed. However, the computed value of $a_3^*(10)$ lies at the upper extreme of the large error bar, and the trend of the observed coefficients is to lower values as ℓ decreases. Similarly the observed values of $a_5(\ell)$ tend to increase more than the computed values at low ℓ-values.

Although the error bars for the a-coefficients at low values of ℓ are large, the trends appear to be real, and, since these ℓ-values sample the deeper regions, the assumption of a rigidly rotating zone just below the convection zone may require revision.

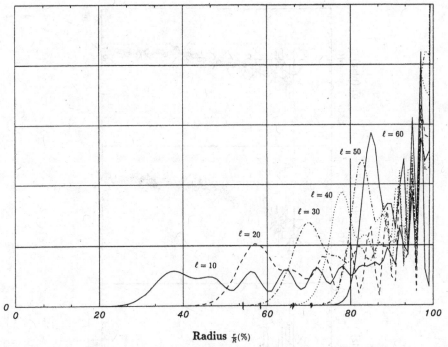

Fig. 12.4 The contribution functions for several values of ℓ are plotted against fractional radius.

12.8 Shell models for the interior rotation

In order to investigate the general internal rotation profile as determined by the splitting coefficients, the solar interior was subdivided into a series of concentric spherical shells chosen in order to optimize the contributions to particular ℓ-values. Although there is some overlap in the peaks in the contribution functions for $\ell = 50$ and 60 (see Figure 12.4), those for $\ell=10$, 20, 30, and 40 are reasonably well separated, and it can be seen that the regions contributing preferentially to the splitting coefficients for $\ell = 10$ and 20 lie in the region $0.3 \leq r/R \leq 0.6$, i.e. well below the estimated base of the convection zone.

The functions Ω_1, Ω_2, and Ω_3 were assigned arbitrary values within these shells, and forward calculations of $a_k^*(\ell)$ were made for $\ell =10$, 15, 20, 30, 40, 50, and 60. The rotation parameters within the shells were varied, and, with a little experience, it was possible to determine values for these parameters within the appropriate shells, such that the theoretical values of the $a_k^*(\ell)$ matched Libbrecht's observed values with an accuracy of a few per cent, well within the error bars.

Theoretical values of $a_k^*(\ell)$ derived for this model (Model 7) are plotted together with the observed values in Figure 12.5, where it can be seen that the theoretical values match the trends in the observed values of a_3^* and a_5^* at low values of ℓ. The corresponding rotation parameters are used to generate appropriate iso-rotation surfaces in Figure 12.6(a), where the surfaces are represented by contour lines on a

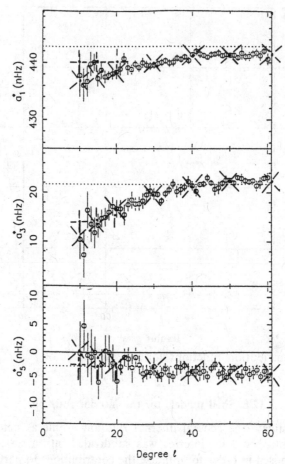

Fig. 12.5 The theoretical splitting coefficients a_k^*, derived from (a) Model 7, represented by '×', and (b) Model 8, represented by '+', are compared with Libbrecht's observed coefficients, represented by 'ϕ'.

section of the Sun through the polar axis. The grey-scale indicates regions of slower-than-average rotation by dark shading and faster-than-average by light shading. Because the discontinuities in the rotation parameters between adjacent shells are artificial, the contours have been smoothed by averaging the parameters over the range $(r/R) \pm 0.02$.

Above $r/R = 0.7$ the rotation profiles are not unlike those derived from Model 6. Initially the angular velocity increases with depth at lower latitudes and then decreases until there is a further rise at $r \approx 0.5$; at higher latitudes the rotation rate decreases initially and then increases, but there is little evidence of a shallow shear zone at the base of the convection zone ($r/R = 0.7$), across which the angular velocity adjusts to that of a rigidly rotating core. The most surprising feature of the

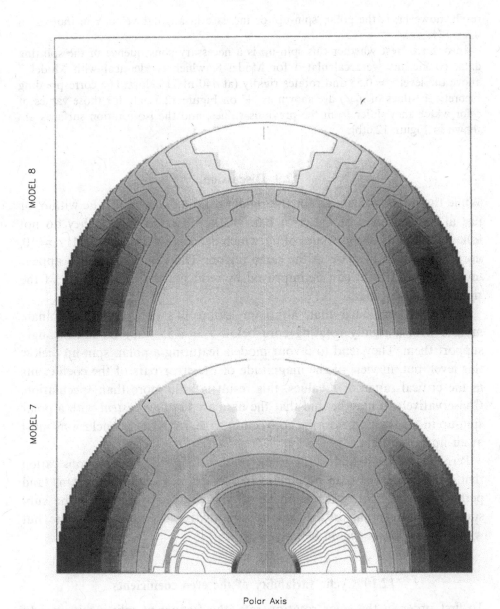

Polar Axis

Fig. 12.6 The iso-rotation surfaces corresponding to (a) Model 7 and (b) Model 8 are represented in meridional section by the contour lines and grey-scale plot. The grey-scale indicates regions of slower-than-average rotation by dark shading and faster-than-average by light shading. The interval between adjacent contours is 15 nHz.

result, however, is the polar 'spin-up' or increase in angular velocity in the region $0.3 \leq r/R \leq 0.4$.

In order to test whether this spin-up is a necessary consequence of the splitting data, coefficients were calculated for Model 8, which is identical with Model 7 above the level $r = 0.55$ and rotates rigidly (at 440 nHz) below. The corresponding theoretical values of $a_k^*(\ell)$ are shown as '+' on Figure 12.5 only for those values of ℓ for which they differ from the previous values, and the iso-rotation surfaces are shown in Figure 12.6(b).

12.9 Discussion

While the points calculated for the rigidly rotating core model lie within, or just at, the extremes of the error bars of the observed values, they do not follow the trends of the values of a_3^*, which decrease sharply towards $\ell = 10$, nor of a_5^*, which increase in the same region. The values of a_1^* also appear too large, but this fit can be improved by reducing the rotation rate of the rigid core to 420 nHz.

It can be concluded that, although Libbrecht's data do not eliminate models with a rigidly rotating core below $r = 0.55$, they do not strongly support them. They tend to favour models featuring a polar 'spin-up' below this level, but, in view of the magnitude of the error bars of the coefficients in the critical range of ℓ-values, this result is little more than speculation. Conservatively, it may be said that the data are *not inconsistent* with a polar spin-up in the region below the convection zone, but firmer conclusions must await improved data.

Perhaps the most important result of this analysis is the demonstration that, when improved data become available, we may obtain important (and perhaps unexpected) information about the detailed structure of the sub-surface rotation profile of the Sun, not only within the convection zone, but within the radiative region below.

12.10 Cyclic variability of the even coefficients

To first order in the solar rotation rate, the frequency splitting is an odd function of m. The even coefficients, therefore, would not normally provide information about rotation and were expected to be negligible or zero. However, non-zero values for these coefficients, which indicate pole–equator asymmetries in the solar structure, have been found by Libbrecht (1989). He realized that the systematic errors likely to be present in the even splitting coefficients should be much smaller than in the odd coefficients and, after examining the values of the even coefficients from earlier data sets as well as his own, found a high degree of internal consistency between the measurements.

Surprisingly, these measurements show a systematic variation between 1982 and 1988. Kuhn (1988) has suggested that they provide an independent measure of cyclic changes in the solar latitude-dependent temperature structure, which may be related to the variations of the cool latitudinal temperature bands observed by Kuhn, Libbrecht, and Dicke (1988), although other mechanisms are also possible. Libbrecht and Woodard (1990) have shown that these bands may be regarded as a proxy for surface activity, which also causes the time dependence of the splittings, a phenomenon which they regard as being due to near-surface perturbations.

The resulting a_2^* and a_4^* coefficients have been calculated by Goode and Kuhn (1990) from both temperature or activity data and the oscillation data. From the latter, they removed the effect of centrifugal distortion from the even coefficients and showed that the a_2^*-coefficients reached a minimum near the activity minimum in 1986, while the a_4^*-coefficients have continued to decline.

Dziembowski and Goode (1991) have re-examined the BBSO data for 1986 and 1988, seeking information about the possibility of detecting toroidal magnetic fields at the base of the convection zone. The only statistically significant results are consistent with a steady, megagauss, quadrupole toroidal field near the base of the convection zone. Such a field is considerably stronger than one would expect from standard dynamo theories. Dziembowski and Goode estimate that the field extends over $\sim 0.1R$ and would therefore contain energy comparable with the luminosity of the Sun through an entire activity cycle. They conclude that, since the solar luminosity changes only slightly during the cycle, this field must be a steady field but warn that confirmation of its existence must await better data.

12.11 Cyclic variability of the odd coefficients

Several authors have looked for time variations in the odd frequency splitting coefficients, in the hope of finding cyclic variations in the Sun's internal rotation profile. Goode and Dziembowski (1991) have studied the equatorial rate at the base of the convection zone and find essentially no variation with the cycle. They do, however, find that the rate at greater depths changes in a way that is anti-correlated with activity, i.e. since 1986 the rate has decreased. Considering the thermal time-scale in the radiative interior and the near constancy of the solar luminosity, they argue that there must be a nearly adiabatic exchange of energy between rotation and magnetic field, i.e.

a torsional oscillation. This possibility was discussed in §7.7 and is further considered in Chapter 15.

Gough and Stark (1992), on the other hand, have examined the BBSO data sets of 1986 and 1988 and found that, on the assumption that rotation rate varies smoothly with radius, both data sets are satisfied (within the published uncertainties) by the same rotation model at all co-latitudes except near $30° - 40°$ and $70°$ and at their southern hemisphere reflections. In agreement with Goode and Dziembowski (1991), they find significant radial trends in the changes to a_2^* and a_4^*, but the evidence for radial trends in the changes to rotationally sensitive combinations of the odd coefficients (see Equations 12.12–12.14) is weak. There are strong anticorrelations between a_2^* and a_4^*, between a_1^* and a_3^*, and between a_3^* and a_5^*, which suggests that the estimates are not independent, and individual coefficients show more evidence of change than do their more 'physical' linear combinations. Gough and Stark consider that the changes noted near colatitude $70°$ might be related to the increase in solar activity between 1986 and 1988.

At present, then, the evidence for cycle-dependent changes in the internal rotation profile is inconclusive. New data from BBSO, covering the period 1986 to 1990, is in preparation and, when it becomes available, we may have a clearer picture regarding this most important aspect of the internal rotation profile of the Sun.

12.12 Conclusion

It has been important to show that the rotational models obtained by inversion methods are not the only models consistent with the current frequency-splitting data. Alternative models which show significant variations across the convection zone are also compatible with these data. In particular, while models exhibiting uniform differential rotation across the convection zone, with strong shears in a shallow transition zone just below it, may not be incompatible with the frequency-splitting data, such models are *not* mandated by these data, and future models of α–ω (or other) dynamos should not be restricted to them. The difficulties regarding the dynamical structure of the convection zone (§12.2) may also be less severe.

As a tool for probing the structure and dynamics of the solar interior, the potential of helioseismology is exciting. However, the accuracy of the currently available data and the methods available for interpreting them have not yet reached the degree of sophistication necessary to resolve the problems which cloud our understanding of the solar dynamo.

Fortunately, helioseismology is a fast-developing field. The GONG Project, a worldwide network of instruments dedicated to maintaining a continuous record of the solar oscillations, is scheduled to yield greatly improved data by 1995. Meanwhile theoreticians and observers are continuing their attempts to refine the data and improve their analytical techniques, and, within the next five years, we may look forward to a greatly improved understanding of the Sun's internal rotation and its relation to the activity cycle.

References

Brown, T. M., and Morrow, C. A.: 1987, *Astrophys. J.*, **314**, L21.

Brown, T. M., Christensen-Dalsgaard, J., Dziembowski, W. A., Goode, P., Gough, D. O., and Morrow, C. A.: 1989, Astrophys. J., **343**, 526–546.

Christensen-Dalsgaard, J., and Gough, D. O.: 1982, *Mon. Not. R. Astr. Soc.*, **198**, 141.

Dziembowski, W. A., and Goode, P. R.: 1991, *Astrophys. J.*, (submitted).

Duvall, T. L. Jr., Harvey, J. W., and Pomerantz, M. A.: 1986, *Nature*, **321**, 500.

Gilman, P. A.: 1983, *Astrophys. J. Suppl.*, **53**, 243.

Gilman, P. A., and Miller, J.: 1986, *Astrophys. J. Suppl.*, **61**, 585.

Gilman, P. A., Morrow, C. A., and DeLuca, E. E.: 1989, *Astrophys. J.*, **338**, 528.

Glatzmaier, G. A.: 1985, *Astrophys. J.*, **291**, 300.

Goode, P. R.: 1992, (Private Communication).

Goode, P. R., and Dziembowski, W. A.: 1991, *Nature*, **349**, 223.

Goode, P. R., Dziembowski, W. A., Korzennik, S. G., and Rhodes, E. J. Jr.: 1991, *Astrophys. J.*, **367**, 649.

Goode, P. R., and Kuhn, J. R.: 1990, *Astrophys. J.*, **356**, 310.

Gough, D. O.: 1984, *Phil. Trans. Roy. Soc. London A*, **313**, 27.

Gough, D. O., and Stark, P. B.: 1992, (Private Communication).

Kuhn, J. R.: 1988, *Astrophys. J.*, **331**, L131.

Kuhn, J. R., Libbrecht, K. G., and Dicke, R. H.: 1988, *Science*, **242**, 908.

Libbrecht, K. G., 1989, *Astrophys. J.*, **336**, 1092–1097.

Libbrecht, K. G., and Woodard, M. F.: 1990, *Nature*, **345**, 779.

Morrow, C. A.: 1988, Ph.D. thesis, University of Colorado.

Parker, E. N.: 1987, *Solar Phys.*, **110**, 11.

Rosner, R., and Weiss, N. O.: 1992, in *Proceedings of the National Solar Observatory/Sacramento Peak 12th Summer Workshop*, ed. K. L. Harvey, San Francisco, California, 511.

Snodgrass, H. B.: 1984, *Solar Phys.*, **94**, 13.

Snodgrass, H. B., and Wilson, P. R.: 1987 *Nature*, **328**, 696.

Unno, W., Osaki, Y., Ando, H., Saio, H., and Sibahashi, H.: 1989, *Nonradial Oscillations of Stars*, University of Tokyo Press.

Wilson, P. R.: 1992, *Astrophys. J.*, **399**, 294.

13

Cyclic activity and chaos†

I believe that order is better than chaos, creation better than destruction,
and, on the whole, I think that knowledge is preferable to ignorance

Kenneth Clark, Civilization

13.1 Introduction

Recent work in the theory of non-linear dynamical systems has centred on
the concept of *chaos*, a term that applies to a great variety of situations and
configurations. This relatively new subject is fascinating in its own right,
and the rapidly growing body of knowledge surrounding it has uncovered
a number of characteristics shared by all *chaotic systems*. The concept has
not only achieved an extensive currency throughout the mathematical and
scientific communities but has also captured the interest of many in the non-
mathematical world, the latter largely due to an excellent popular discussion
of the history and basic ideas by James Gleick (1988).

If the dynamics of the solar magnetic field are due to magnetoconvective
dynamo action within the Sun, then the activity cycle is governed by the
non-linear equations of magnetohydrodynamics, discussed in Chapter 11. A
number of investigators have suggested that solar and stellar activity cycles
are chaotic phenomena and have begun to explore the implications of cyclic
systems which are chaotic.

If stellar activity cycles are indeed examples of chaotic systems, then they
will share in the universal characteristics of such systems. In order to discuss
the implications for cyclic activity, a brief outline of the relevant concepts in
the theory of chaos is called for. The interested reader is advised to consult
the literature for a fuller account; in addition to that by Gleick (1988), more

† This chapter was researched and drafted by Herschel B. Snodgrass.

228

technical discussions abound: see, e.g. Baker and Gollub (1990) or May (1976).

13.2 Definition

A fully chaotic dynamical system is one which, although in principle deterministic, has such extreme sensitivity to initial conditions that its behaviour is, in practice, unpredictable. The system's behaviour would be predictable only if the initial conditions could be prescribed with infinite accuracy; any infinitesimal difference in the initial conditions diverges *exponentially* in time.

This remarkable behaviour is characteristic of dynamical systems for which the governing equations are non-linear. Although sharing apparent similarities, chaotic systems differ fundamentally from those which are not deterministic, such as a narrowband Gaussian oscillator, where a deterministic system is subjected to stochastic perturbations.

Central to the description of chaotic behavior is the notion of phase space. Through the introduction of suitable variables, $x_i\{i = 1, n\}$, which may represent the components of velocity and acceleration as well as position, the equations governing the system are converted into first-order ordinary differential equations in time, usually expressible in the Hamiltonian form

$$\frac{dx_i}{dt} = F_i(x_j, \ldots, f_k, \ldots,) \quad \{j = 1, \ldots, n, k = 1, \ldots, m\} \qquad (13.1)$$

for each i, where f_k represents the forces or controls acting on the system. The space spanned by the variables x_i is called the *phase space* of the system. As the system evolves in time, the system point $[x_1(t), x_2(t), \ldots, x_n(t)]$ moves about in this n-dimensional space along a path called the *system trajectory*. Thus, for example, the phase space of a one-dimensional oscillator has two dimensions, the coordinates being position $x(t)$ and velocity $\dot{x}(t)$, while the dynamical condition is specified by $\ddot{x} = -kx$. By defining $x_1 = x$, $x_2 = \dot{x}_1$, and $f = -kx$, Equation (13.1) becomes

$$\frac{dx_i}{dt} \quad \{i = 1, 2\} = \{x_2, -kx_1\}.$$

If there is no frictional damping, the trajectories are ellipses centred on the origin whose axes are determined by the initial position and velocity, $x_1(0)$ and $x_2(0)$.

There are many important properties of these system trajectories, the most important of which, for this brief description, is that they never cross: i.e. that no point in phase space lies on more than one system trajectory, since the motion is uniquely determined by the values of the system variables at any time.

13.3 The onset of chaotic behaviour

Three concepts in the general theory of chaotic systems are particularly relevant here. The first concerns the onset of chaotic behaviour in simple non-linear systems. As one adjusts certain parameters in the non-linear equations, the behaviour of the system changes. For some ranges of the parameters, the phase-space trajectory is closed and therefore predictable, perhaps settling down to a specific point or line which may be associated with the initial conditions, or oscillating, in which case the trajectory is a closed curve or fills a portion, or hypersurface, of the phase space uniformly. For other ranges of the parameters, phase-space trajectories which are neighbouring, initially, diverge exponentially in time, so that the topology of the system's phase space is no longer simple, and the system is no longer predictable; it has become chaotic.

The transition from predictable to chaotic behavior, as one or more critical parameters are changed, is generally characterized by a succession of bifurcations at which increasingly complex, multiply periodic motions occur. A remarkable theorem states that if, for some range of the parameters, the system exhibits a multiply periodic motion with three distinct frequencies (York 1975), then there will also be a range of parameters for which the motion is chaotic.

An illustration of this is provided by the logistic difference equation, which arises as a simple model of population growth in a limited environment from one generation, Y_i, to the next, Y_{i+1},:

$$Y_{i+1} = \lambda Y_i(1 - Y_i). \tag{13.2}$$

This apparently simple equation has some peculiar properties, which are discussed in detail by May (1976). Here we shall highlight the main features. A system with behaviour governed by Equation (13.2) exhibits different sorts of behaviour for different ranges of the parameter λ. Obviously Y_{i+1} cannot be negative, so the values of Y_i must lie in the range $0 \le Y_i < 1$. Further, since the maximum value of $Y_i(1 - Y_i)$ is $\frac{1}{4}$, which occurs at $Y_i = \frac{1}{2}$, λ cannot exceed 4.

Starting from any initial value Y_0, in the range $0 < Y_0 < 1$, with λ in the range $0 < \lambda < 4$, the sequence of iterations soon loses track of its initial value. For $\lambda \le 1$ it converges to zero, while for $1 < \lambda \le 3$ it converges to the limit $Y_\infty = 1 - (1/\lambda)$. As λ increases from 3 to 4, the values of Y_i at first cycle regularly between two values, then four, then eight, and eventually without the repetition of any discernible pattern. This is illustrated in Figure 13.1(a), where all the values of Y_i that are obtained after a very large number of iterations are plotted at each value of λ.

Examples of the sequences, Y_i, obtained for three different values of λ are illustrated in Figure 13.1(b). For $\lambda = 2.98$ (i.e. just before the first bifurcation shown in Figure 13.1(a)) the sequence converges to a single value; $\lambda = 3.47$ occurs just to the right of the second bifurcation shown in Figure 13.1(a), and the sequence jumps between four values in an endlessly repeated pattern; $\lambda = 3.6$ occurs after many bifurcations have taken place, and the sequence passes through many values without discernible patterns appearing.

To see how this works, note that, if one starts with $Y_0 = 1 - (1/\lambda)$, then $Y_i = Y_0$ for all n, giving an exact solution for all values of $\lambda > 1$. Now perturb this initial value by a small amount ϵ: $Y_0^* = Y_0 + \epsilon$, and iterate to obtain, to first order, $Y_1^* = Y_0 + (2 - \lambda)\epsilon$. Convergence to $1 - (1/\lambda)$ will occur if this new departure is smaller in absolute value than ϵ, i.e. provided $1 < \lambda < 3$. This is the point where the first bifurcation (see Figure 13.1(a)) sets in. Further analysis (see eg. Baker and Gollub 1990) shows that at $\lambda \approx 3.56$ the number of bifurcations approaches infinity, and the system slides into chaos. Beyond this value of λ, there are other ranges in which the behaviour briefly returns to alternations between a few discrete values and then becomes chaotic again, as seen in Figure 13.1(a).

13.4 The attractor

The second important concept is that of an *attractor*. For a dissipative system, the phase space volume associated with a family of neighbouring trajectories decreases in time. That is, a dissipative system will 'settle down' to a region of phase space that is smaller, and generally of lower dimension, than the full space available to it through the range of its possible initial conditions. This region, or subspace, of the phase space is called an 'attractor'. Possible attractors of an ordinary kind can be illustrated through a few examples: an exponentially damped non-oscillatory motion settles down asymptotically to a point (Figure 13.2(a)); a two-dimensional oscillator, oscillating in the x, y plane, subject to a damping force in the x-direction alone, settles down to an oscillation along the y-axis, i.e. to a phase orbit in the $y, \dot{y}$ phase plane. For a certain range of parameters, a simple harmonic oscillator, damped in both the x and y directions and subject to a sinusoidal driving force, settles down to a closed orbit (the *limit cycle*, Figure 13.2(b)) in phase space.

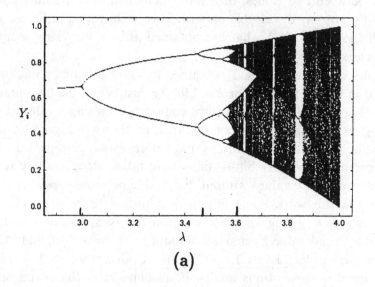

(a)

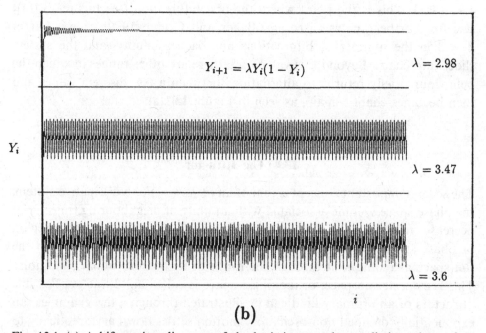

$$Y_{i+1} = \lambda Y_i(1 - Y_i)$$

$\lambda = 2.98$

$\lambda = 3.47$

$\lambda = 3.6$

(b)

Fig. 13.1 (a) A bifurcation diagram of the logistic map shows all the values of Y_i obtained after a large number of iterations at each value of λ. (b) Examples of the sequences, Y_i, obtained for three different values of λ. For $\lambda = 2.98$ (i.e. just before the first bifurcation shown in (a)) the sequence converges to a single value; $\lambda = 3.47$ occurs just after the second bifurcation shown in (a), and the sequence jumps between four values in an endlessly repeated pattern; $\lambda = 3.6$ occurs after many bifurcations have taken place, and the sequence passes through many values without discernible patterns appearing.

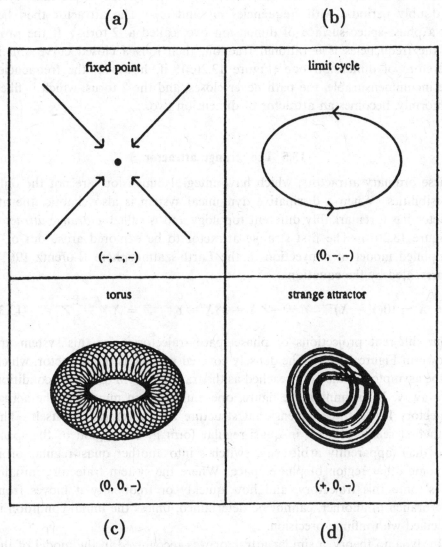

Fig. 13.2 The diagram illustrates four different attractors; the point, the limit cycle, the torus, and the strange attractor.

The *dimension* of an attractor, which is the third important concept, is, roughly speaking, the dimension of the region of phase space that it occupies. Thus, in the examples of the previous paragraph, the *point* is an attractor of dimension zero, and the *closed curve* is an attractor of dimension one. A phase space trajectory, being a line, is, of course, one-dimensional. One can get attractors of higher dimension, however, by filling a portion (hypersurface) of the phase space with a trajectory that does not close on itself. For example, consider the trajectory of the system whose limit cycle is doubly periodic with frequencies ω_1 and ω_2. The attractor then lies on a phase-space surface of dimension two, called a '2-torus'. If the ratio of the frequencies is a rational fraction, then it is a closed curve and is therefore of dimension one (Figure 13.2(c)); if, however, the frequencies are incommensurable, the path never closes, and the 2-torus, which is filled uniformly, becomes an attractor of dimension two.

13.5 The strange attractor

These ordinary attractors, which have integral dimension, are not the only possibilities. When a dissipative dynamical system is also chaotic, the attractor has a remarkably different topology and is called a *strange attractor* (Figure 13.2(d)). The first strange attractor to be explored arose out of a simplified model of convection in the Earth's atmosphere (Lorentz 1963), represented by the equations:

$$\dot{X} = 10(Y - X); \quad \dot{Y} = -ZX + 28X - Y; \quad \dot{Z} = XY - \frac{8}{3}Z. \quad (13.3)$$

Four different projections of phase-space trajectories for this system are shown in Figure 13.3(a). The densely covered portion is the attractor, which is the asymptotic trajectory reached as the transient from the initial condition dies away. In examining the figure, one must bear in mind that the actual trajectory has a three-dimensional structure and never crosses itself. The path first seems to orbit in quasi-regular form in one region of the space and then, apparently arbitrarily, switches into another quasi-regular orbit in some other region of phase space. Where the system trajectory initially 'falls' onto this trajectory, and how quickly or frequently it moves from one branch to another, cannot be determined, unless the initial condition is specified with infinite precision.

In dynamo theory, a similar attractor was recognized in the model of the *coupled disk dynamo* (Cook and Roberts 1970), for which the equations are:

$$\dot{X} + \mu X = (A + V)Y, \quad \dot{Y} + \mu Y = VX, \quad \dot{V} = 1 - XY, \quad (13.4)$$

where A is a constant, X and Y represent currents, and V represents the angular velocity, all in dimensionless form. Four different projections of the resulting strange attractor are seen in Figure 13.3(b).

Other examples abound, and they all have the peculiar property of jumping in an unpredictable way from one region of phase space to another. This property of the topology of a strange attractor can be summarized through

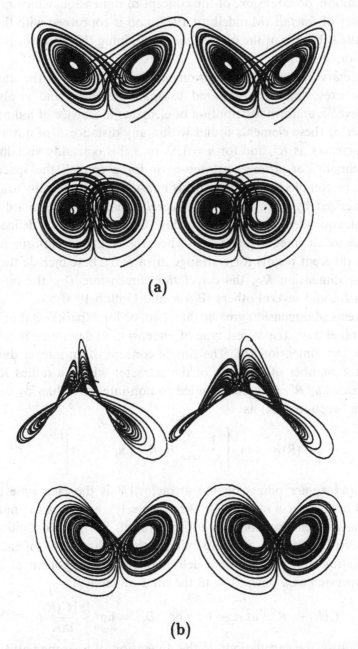

Fig. 13.3 (a) Four different projections of the strange attractor arising from the Lorentz model of convection in the Earth's atmosphere. (b) Four different projections of the coupled disk dynamo strange attractor.

a generalization, or extension, of the concept of dimension, which originated in the theory of *fractals* (Mandelbrot 1977) and is concerned with the degree to which the subspace of the phase space containing the attractor is filled by the attractor.

The ordinary concept of dimension refers, of course, to the number of coordinate axes which are required to span the space and is always an integer. Given a uniform distribution of elements in a space of n-dimensions, the number of these elements found within any distance R of any reference point O increases as R^n, and for $n = 1, 2,$ or 3 this is readily visualized.

For a chaotic, or strange, attractor, or for a fractal, the space is not uniformly, or completely, filled, and this non-uniform (or 'frothy') filling can be characterized by giving it a *non-integral* dimension. There are actually several non-equivalent ways in which this 'dimension' can be defined, all of which lead to an integer dimension when the filling is uniform but yield somewhat different results for a strange attractor. These include the fractal, or *capacity* dimension D_C, the *correlation* dimension, D_G, the *information* dimension D_I, and several others (Baker and Gollub 1990).

For systems of dimension greater than two, or for experimental or numerically generated data, the easiest type of dimension to determine in practice is the correlation dimension D_G. The simple concept of an integer dimension, in which the number of elements of the attractor within a radius R of that point increases as R^n, may be extended to non-integral values by defining a correlation function $C(R)$ as

$$C(R) = \lim_{N \to \infty} \left[\frac{1}{N^2} \sum_{i,j=1}^{N} H(R - |\mathbf{x}_i - \mathbf{x}_j|) \right], \tag{13.5}$$

where $\mathbf{x}_i$ and $\mathbf{x}_j$ are points on the attractor, H is the Heaviside function ($H(y) = 0$ or 1 when $y < 0$ or $y \geq 0$ respectively), and N is the number of points randomly chosen from the entire data set. With this definition, $C(R)$ gives the average fraction of points lying within a distance R of each other. The correlation dimension is then determined as the exponent of R, when $C(R)$ is expressed as a power law in the limit of small R:

$$C(R) \to R^{D_G} \text{ as } R \to 0; \quad \text{or} \quad D_G = \lim_{R \to 0} \frac{\ln\{C(R)\}}{\ln R}. \tag{13.6}$$

The details of the calculations of the dimension of a strange attractor are discussed by (e.g.) Packard *et al.* (1980), and by Grassberger and Procaccio (1983). In practice, they depend on the availability of a suitably large number of data points, N, separated by a suitable time interval τ and, since N must be of the order of a few thousand points per phase-space dimension, this raises considerable practical difficulties for the investigation of some situations.

In summary, then, if one begins with the set of dynamical equations for a dissipative system, written in Hamiltonian form, one finds by adjusting the parameters in these equations that there are distinct regimes of behaviour. For some values of the parameters, the behaviour of the system may be predictable, and for others chaotic. The transitions between these extremes are characterized by bifurcations at which the system alternates between increasing numbers of well-defined but distinct regions of phase space, in which the system exhibits multiple periodic oscillations. Each regime is characterized by an attractor (for some systems there are actually several *basins of attraction*, that is, not just one but several possible disjoint attractors). The dimension and character of the system's attractors therefore vary as the system parameters are changed, and, in regimes where the behaviour is chaotic, the attractor is a strange attractor and has a non-integral dimension.

The determination that the behaviour of a system is chaotic is important, since it means that the system is guaranteed to be unpredictable in the long term. That is, even if one knows all about the strange attractor for a system, as one does, for example, for the Lorentz system (cf. Figure 13.3(a)), one cannot determine precisely where the system point lies on this attractor, or at which point it will land on the attractor for given initial conditions, and one cannot therefore predict its subsequent trajectory. Typically, it may continue in its quasi-periodic orbit in one of the phase-space regions for some time and then move off to the other one. But exactly when this will happen can only be known if one knows *precisely* where it is in the first place, information which is rarely available to the required accuracy. While the realization that the system in which one is interested is chaotic may be unwelcome for those concerned with long-term predictions, it is probably desirable to face reality. Even for chaotic systems, short-term predictions of some reliability may be possible while the system is in a quasi-periodic orbit in the phase space of a strange attractor, but such predictions are always subject to the uncertainty of the system moving off this attractor.

13.6 Chaos and dynamo theory

One example of a non-linear dissipative system with chaotic regimes is the coupled disk dynamo, defined by Equation (13.4). Although quite interesting in its own right, this system probably has little relevance to the dynamos thought to be present in the Sun and solar-type stars. Other simplified systems that are more relevant to solar and stellar activity cycles have, however, been investigated, and one of these is described here: a parameterized mean-field $\alpha-\omega$ dynamo in one space dimension, made non-linear through the Lorentz force interaction between the magnetic field and the differential rotation. Jones, Weiss, and Cattaneo (1985) have studied this system, determined the regimes of regular and chaotic behaviour, displayed the corresponding attractors, and compared the activity behaviour with that of the Sun.

The dynamo number D, which is a measure of the 'effectiveness' of the dynamo (see § 11.3), is a stability parameter for this system, and the different regimes of behaviour occur as its value, defined by Equation (11.20), is changed.

Jones, Weiss, and Cattaneo have studied a sixth-order system which is a complex generalization of the Lorentz equations (Equations (13.3)) and found behaviour which follows the anticipated pattern. There are exact periodic solutions for $D > 1$ which vary sinusoidally with time. As D increases above 2, the solutions become at first doubly periodic, then triply periodic for $D \sim 3.5$, and then slide into a cascade of bifurcations which lead to chaos for $D \geq 3.8$. The chaotic patterns are, however, interspersed with windows in which regimes of quasi-periodicity are interrupted by episodes of reduced activity. In this feature, the behaviour of the system is not unlike the grand minima in the observed patterns of solar activity.

The model described by these equations is the 'simplest' model that captures the essential physics of a non-linear stellar dynamo (Cattaneo, Jones, and Weiss 1983). It should be noted, however, that the system contains a term equivalent to the torsional oscillations, which therefore play a critical role in this model by supplying the non-linear effect that allows chaotic behaviour to occur. A chaotic $\alpha-\omega$ dynamo has been investigated by Schmalz and Stix (1991).

13.7 Is the sunspot cycle chaotic?

What can be learned from these investigations? The magnetohydrodynamic equations which are believed to govern the magnetic activity of the Sun are certainly capable of producing chaotic behaviour. Further, the resemblance between the actual sunspot cycle (Figures 5.1 and 5.4), with its grand minima interspersed with aperiodic oscillations, and the chaotic activity cycle produced by the simplified model discussed in the preceding section, suggests that the solar cycle may indeed be a chaotic phenomenon. One must recall, however, that this behaviour is also consistent with the model of a randomly perturbed oscillator (the narrowband gaussian process).

It has been pointed out by Ruzmaikin (1990) that the solar cycle may be regarded as random on short time-scales, i.e. $\tau \leq 1$ Carrington Rotation (the time-scales for the eruptions and lifetimes of individual active regions), but may be chaotic on longer time-scales, i.e. $\tau \geq 1$ year.

To show that the activity cycle is chaotic it is necessary to show that the solar attractor is strange, i.e. that it has a non-integral dimension. The question then is whether or not we have enough data, or the appropriate kind of data, on hand to determine the dimension of the cycle. As shown in Chapter 5, the sunspot number $R_Z = N(t)$ must be regarded as a sequence N_i, with t_i a discrete rather than a continuous variable. The longest reliable time-series available is provided by the annual sunspot numbers, which span approximately 300 years since 1703. Thus, the total number of data points, N_i, available for this calculation is of the order of 300, of which only those obtained since 1849 are reliable in detail. In view of the above discussion, this number seems completely inadequate. Since the dimension is expected to be $D_G > 3$, one would anticipate the need for $\sim 10^4$ points. Furthermore, very little data is available from even the most recent of the grand minima (the Maunder Minimum), and the determination of whether or not the cycle is chaotic must certainly take such periods into account.

In spite of this situation, a number of workers have attempted to calculate the attractor dimension from the data on hand. Makarenko and Ajmanova (1988) have examined the sunspot number records and estimate a dimension close to 2. Suess (1965, 1978) has compiled an extensive proxy data set extending over 6000 years, using the abundance of ^{14}C in tree ring data and, from these data, Gizzatulina *et al.* (1990) estimate a dimension closer to, but less than, 3. These results imply that the attractor associated with the solar cycle is indeed fractal and that the associated dynamical system is chaotic. However, J. Thomas (private communication) has pointed out that the use of proxy data inevitably introduces another physical system into the problem, and it may be that the fractal dimension so derived reflects the nature of the proxy system rather than the one under investigation.

Nevertheless, because these estimates of the fractal dimension of the solar attractor are low, the dynamical system is likely to be a simple one, which can be described in terms of only a few modes. Thus Ruzmaikin (1990) has described a two-parameter limit cycle in terms of mean field dynamo theory (see Chapter 11). Gudzenko and Chertopud (1978) have attempted to model solar cyclic activity by a second-order system which produces limit cycles, with the idea that the aperiodicity is the result of a noise source external to the cyclic process. Although they find an aperiodicity, it is quite unlike the Maunder intermission. Fifth-order systems, such as that described by Childress and Fautrell (1982), can be richer in that kind of behaviour, though the fifth-order system of Cattaneo, Jones, and Weiss (1983) shows no intermission periods.

Spiegel and Weiss (1980, see also Spiegel 1985) have proposed a simple version of the solar activity cycle in which the strong fields are formed immediately below the convection zone, from which they escape cyclically through the agency of an instability, such as buoyancy. They have experimented with a fifth-order system which is a generalization of the Lorentz system and, for a wide range of parameter values, obtain time-series showing mild chaos and intermissions of moderate length, but none which greatly resemble the solar cycle. They find that a signed parameter, which they associate with magnetic field, generally changes sign from cycle to cycle, but sometimes it does not, and Spiegel cautions that, just because we have seen a few reversals of the solar poloidal field, this does not mean that it is necessarily a permanent effect. Spiegel prefers to regard the cycle as an $\sim$ 11-year phenomenon, subject to aperiodic or chaotic excursions in both amplitude and sign, and believes that he is seeing a strange attractor of dimension less than five.

In some recent work Durrant (private communication) has investigated the stability, or convergence, of these calculations and found that, as expected, the number of data points (288) is far too small for the extraction of the dimension D_G by the method outlined in § 13.4. Durrant has also applied his analysis to the varve data (see § 5.5), for which he was able to obtain $\sim$ 1600 points; but, once again, he found no sign of convergence. The implication of this is that, even if the varves had been a true proxy for a chaotic system, the number of data points still would have been too small to permit the calculation of the correlation dimension for the activity cycle attractor.

Finally, Durrant has used Equation (13.6) to find the dimension of the attractor generated by the sixth-order system of Jones, Weiss, and Cattaneo. The number of cycles that can be generated is now arbitrarily large, and the results obtained show excellent convergence. From this calculation Durrant finds, using $\sim$ 9600 data points, that, for a dynamo number $D = 3.0$, for which the system is doubly periodic, the dimension is $D_G \approx 2.1$. As the dimension should be integral, (i.e. equal to 2 in this case), the accuracy of this method may be assessed. For $D = 8$ on the other hand, the system is chaotic; Durrant finds $D_G \approx 3.4$, while Jones, Weiss, and Cattaneo have estimated that this value should be less than 5.

13.8 Conclusion

Although chaos theory has not yet reached the stage at which the identification of chaotic behaviour can lead to an understanding of the physics underlying the system, it does offer an important classification or characterization of the system by comparison with other chaotic systems.

In the case of the solar cycle, chaos theory emphasizes the value of the apparently elementary one-dimensional properties of the cycle. Unfortunately, attempts to identify those regions in phase space with strange attractors associated with the quasi-periodic cycle and intermissions typified by the Maunder Minimum period are hampered by the limited span of the data string provided by the Zurich sunspot numbers, or even of the more extensive ^{14}C tree ring data.

It thus appears highly unlikely that the question of whether the solar activity cycle is a chaotic system can be resolved in the foreseeable future, unless some reliable proxy data can be found. This was one of the reasons why the alleged 'varves' (cf. §5.5 above) generated so much excitement, and it should certainly provide an impetus for the continuing search for some reliable proxy data, such as might be provided by, for example, ice layerings in Antarctica or Greenland. Proxy data, however, are always subject to the caveat that, unless the physics of the proxy system are well understood, the fractal dimension so determined may reflect the proxy system rather than the system under study.

If, indeed, the solar cycle is chaotic, then long-term forecasting of the properties of individual cycles is not feasible. This does not mean, however, that one would not be able to forecast from one cycle to the next, or to anticipate the coming of a grand minimum from a few years ahead. If the attractor were thoroughly mapped out, the switching from one region of the attractor to another might become apparent before the patterns of magnetic activity on the Sun associated with the new region of the attractor were to appear. The problem of forecasting the solar activity cycle is further discussed in the next chapter.

References

Baker, G. L., and Gollub, J. P.: 1990, *Chaotic Dynamics, an Introduction*, Cambridge University Press.

Cattaneo, F., Jones, C. A., and Weiss, N. O.: 1983, in *Solar and Stellar Magnetic Fields, Origins and Coronal Effects*, ed. J. O. Stenflo, 307.

Childress, S., and Fautrell, Y.: 1982, *Geophys. and Astrophys. Fluid Dyn.*, **22**, 235.

Cook, A. E., and Roberts, P. H.: 1970, *Proc. Camb. Phil. Soc.*, **68**, 547.

Gleick, J.: 1988, *Chaos*, Viking Penguin, New York.

Grassberger, P., and Procaccio, I.: 1983, *Physica*, **9D**, 189.

Gizzatulina, S. M., Ruzmaikin, A. A., Rukavishnikov, V. D., and Tavastsherna, K. S.: 1990, *Solar Phys.*, **127**, 281.

Gudzenko, L. L., and Chertopud, V. E.: 1964, *Soviet Astron.*, **41**, 697.

Jones, C. A., Weiss, N. O. and Cattaneo, F.: 1985, *Physica*, **14D**, 161.

Lorentz, E. N.: 1963, *J. Atmos. Sci.*, **20**, 130.

Makarenko, N. G., and Ajmanova, G. K.: 1988, *Astron. Tsirk.*, **1533**, 21.

Mandelbrot, B. B.: 1977, *Fractals: Form Chance and Dimension*, W. H. Freeman and Co., San Francisco, California.

Packard, N. H., Crutchfield, J. P., Farmer, D. J., and Shaw, R. S.: 1980, *Phys. Rev. Lett.*, **45**, 712.

May, R. M.: 1976, *Nature*, **261**, 459–67.

Ruzmaikin, A. A.: 1990, *Solar Photosphere, Structure, Convection and Magnetic Fields*, ed. J. O. Stenflo, I.A.U. Symposium, 343.

Schmalz, S., and Stix, M.: 1991, *Astron. Astrophys.*, **245**, 654.

Suess, H. E.: 1965, *J. Geophysics Res.*, **70**, 5937.

Suess, H. E.: 1978, *Radiocarbon*, **20**, 1.

Spiegel, E. A.: 1985, in *Chaos in Astrophysics*, eds. R. J. Buchler, J. M. Perdang, and E. A. Spiegel, D. Reidel, Dordrecht, 91.

Spiegel, A. E., and Weiss, N. O.: 1980, *Nature*, **287**, 616.

York, J. A.: 1975, *American Math. Monthly*, **82**, 985.

14

Forecasting the solar cycle

It's tough to make a prediction, especially about the future
                                                        Yogi Berra

14.1 Introduction

Chapter 1 discussed the relationship of the solar cycle to the terrestrial environment and offered the hope that a greater understanding of solar and stellar cycles might lead to improved predictions of the parameters of a given solar cycle and, consequently, more reliable forecasts of solar-induced geospheric phenomena.

The more important terrestrial consequences arise during and just after sunspot maximum, when the EUV–UV and total solar irradiance are also at a maximum, along with the probability of occurrence of large solar flares, particle bursts, and solar cosmic rays. These enhanced solar emissions disrupt communications, shorten the orbital lifetimes of satellites in low Earth orbit, cause failure in solid-state components in satellites, and generally introduce major or minor disruptions to our environment (see § 1.3). The likelihood of the occurrence of disruptive events follows closely the intensity of the activity cycle, so that accurate forecasting of solar activity on time scales of weeks, months, and years is of considerable importance to those agencies which plan and operate space missions.

Apart from space missions, other possible terrestrial effects of the influence of solar activity are discussed in Chapter 1 and elsewhere in this book. Perhaps the most significant for us and our descendants are the small variations of the solar output which accompany the activity cycle. Although the measured variations are only a few tenths of a per cent, it is generally accepted that, because of the non-linear nature of the interaction of solar radiation with the geosphere (which is poorly understood), the effect of these small fluctuations on our environment may be considerably amplified. If, indeed, the climatic excursions which occurred during the Maunder Minimum were causally related to a somewhat greater reduction of the Sun's output, then our need to understand and, if possible, predict the likely occurrence of future grand minima assumes considerable significance. To the time-scales of weeks, months, and years, mentioned in connection with space missions, it is therefore appropriate to add decades and even centuries.

Not only do NASA and other US and European government agencies invest a great deal of funds in solar research in the hope of improving solar-terrestrial forecasting, but almost every grant application for support for research into all areas of solar physics also includes at least one paragraph extolling the advantages of improved forecasting, arising from a greater understanding of solar processes. It is time now to assess the pay-off.

14.2 The tactics and strategies of forecasting

In any type of forecasting, be it picking winners in horse-racing events, playing the stock market, or forecasting the weather or the sunspot cycle, four general strategies may be identified: (i) random guesswork, (ii) numerology or statistical methods, (iii) pattern recognition techniques, and (iv) physical modelling. The first, although often used by amateur punters and stockmarket novices, is not to be recommended for the serious scientist. The second has its faithful devotees among racegoers who follow winning numbers, barrier positions, or the almost infinite variety of 'systems' which take account of the numerical records of the contestants in a racing event without regard to the physical condition of the horse, the jockey, or the course. In the stock market, the pattern recognition technique is utilized by the chartists who identify trends in the price history of a stock in terms of 'rounded bottoms', 'head and shoulders tops', and many other exotic patterns, without reference to the basic operations of the company concerned or the prevailing economic conditions.

In sunspot forecasting, the devotees of the numerological or pattern recognition approaches search for multiple periodicities and amplitudes in the sunspot or other data records and attempt to combine these in a form which is consistent with past cycles. The advantage of these techniques is that they permit the forecasting of many future cycles, since they assume that all the important periodicities have already appeared in the data. The disadvantage is that no physical information is used and that if, as discussed in Chapter 13, the activity cycle is best described in terms of the strange attractors of a chaotic system, then the technique is, by definition, useless, since no strict periodicities exist and arbitrarily close initial conditions may yield divergent trajectories. Predictions of the amplitude of Cycle 22 by this method have not, in fact, fared particularly well, since most of them forecast low values of R_{Z_M} (Brown 1988) which were exceeded early in the cycle.

Since this text has been devoted to achieving some physical understanding of stellar cyclic behaviour, it is scarcely surprising that it leans towards the fourth forecasting strategy, that of attempting to base predictions on physical rather than numerical models of the cycle.

14.3 Forecasts based on physical models

Most of the physical modelling techniques which have been applied to date are surprisingly primitive, being based on the elementary Babcock model, in which the polar poloidal field at minimum is wound into toroidal bands which erupt to form the surface active regions. Although this model is now regarded as mainly of historic interest (see Chapter 7), many forecasters, e.g. Schatten *et al.* (1978), have based their predictions on the assumption that the strength of the polar fields at sunspot minimum is causally related to the sunspot fields at maximum and should therefore provide a reliable basis for forecasting.

Unfortunately, since the polar fields are weak, even at sunspot minimum, and are approximately transverse to the line of sight, direct measurements of these fields are difficult to obtain, even at the present time, and are unavailable for cycles prior to Cycle 21. Synoptic magnetic charts have nevertheless been prepared by the Mount Wilson Observatory (MWO) since 1967 and by the Wilcox Solar Observatory (WSO) since 1976, and, from these data, Layden *et al.* (1991) have constructed plots of the polar fields (at both poles and from both observatories) as a function of time. These plots show annual variations due to secular changes in the Earth's heliographic latitude as it orbits the Sun, and the true polar field variations are represented by the envelope of these curves.

Since the scatter in the MWO data is considerably greater than that in the WSO data, particularly prior to 1980, the WSO data are to be preferred. These show that, at the minimum prior to Cycle 22, the strength of the polar field, $B_P(\min)$, was about 30% stronger than for Cycle 21. The assumption of direct proportionality between $B_P(\min)$ and the Zurich sunspot number at maximum, $R_Z(\max)$, yields an estimate for $R_Z(\max)$ for Cycle 22 of 215, a value which proved to be too large. However, the value of $B_P(\min)$ for Cycle 21 may be an underestimate, since it may have peaked in the year before

WSO began making observations. Layden *et al.* conservatively estimated that $R_Z(\text{max})$ should lie in the range 190–215 (or less).

Schatten and Hedin (1984) used direct polar field measurements from MWO to predict an $R_Z(\text{max})$ for Cycle 22 of 110 ± 10. After the cycle had begun, Schatten and Sofia (1987), with the benefit of a little hindsight, gave several reasons why that prediction was too low and, using the WSO data, offered a prediction of 180 (with no error estimates).

Predictions based on only two data points are obviously hazardous, and too much importance should not be placed on these results. However, when discussing the proxy data for the polar fields, Layden *et al.* (1991) argue cogently for the inclusion of a zero-point term in the linear relationship between the measure of solar activity at maximum, A, with the polar field strength at minimum, B, of the form

$$A = m(B - \gamma), \tag{14.1}$$

with γ representing the threshhold polar field and m a constant of proportionality.

In the spirit of the Babcock model, they argue that, because the toroidal field must reach some critical density before the instabilities which give rise to active regions can develop, there may be some threshhold level for the polar fields below which the normal cyclic activity cannot occur. They suggest that polar fields which only marginally exceeded the threshhold should require a longer winding time to reach the critical magnetic densities in the toroidal fields, while strong polar fields should require a shorter winding time, a scenario which is consistent with the relatively shorter rise times of the stronger cycles. Layden *et al.* also argue that the low activity of the Maunder Minimum was due to the polar fields falling below the threshhold for some reason.

Since the Babcock model is today not considered to provide a plausible physical picture of the cycle, the premise on which these arguments are based is questionable. Further, if Equation (14.1) were applied directly to the polar fields, the estimates of $R_Z(\text{max})$ for Cycle 22 based on the polar fields would have to be increased significantly. Any fair assessment of the value of $B_P(\text{min})$ as a precursor predictor of $R_Z(\text{max})$ (with or without regard to the Babcock model) must, however, wait until the latter part of Cycle 23, when a second set of direct measures of $B_P(\text{min})$ may be compared with $R_Z(\text{max})$.

14.4 Precursors and proxies

A half-way step between the numerology approach to forecasting and physical modelling methods is the use of *precursors*. This technique is statistical in nature but is based on correlating some solar or geophysical property which can be observed during or prior to the minimum phase of the old cycle with a representation of the amplitude of the following cycle.

Whenever a good correlation is obtained, the precursor can be used for forecasting and should initiate an investigation of the possibility of a physical link between the correlated parameters. Some possible precursors are (i) the sunspot number at sunspot minimum $R_Z(min)$; (ii) the initial growth rate parameters of the new cycle; (iii) the shape and structure of the corona at minimum (including the polar field bending angle); (iv) the number and latitude distribution of the polar faculae; (v) the structure of the interplanetary magnetic field; and (vi) the geomagnetic indices aa and Ap, which measure the response of the geomagnetic field to fluctuations in the solar wind.

It has been argued (e.g. Layden *et al.*, 1991) that many of these precursors (e.g. (iii)-(vi)) are simply *proxies* for $B_P(min)$, the polar magnetic field at sunspot minimum, and that therefore their application to forecasting should be regarded as physical modelling rather than numerology. Because reliable data connecting any of these quantities with $B_P(min)$ is slight (owing to the lack of an extensive data base for $B_P(min)$), the author regards such claims with some caution and prefers to consider the value of any of these phenomena as precursors purely with regard to the availability of a reliable data base extending over a sufficient number of cycles to be able to establish a correlation between $P_c(min)$, the value of the precursor at (or before) sunspot minimum, and $R_Z(max)$.

14.4.1 The sunspot number at sunspot minimum

The quantity for which the most extensive data base is available is $R_Z(min)$. Using simple linear regression analysis for estimating $R_Z(max)$, R. M. Wilson (1990b) derived the relation

$$R_Z(max) = 81.7 + 5.46R_Z(min),$$

which yields a Pearson correlation coefficient of only 0.45, and he concludes that $R_Z(min)$ is not a reliable indicator of $R_Z(max)$. Interestingly, his estimate of $R_Z(max)$ for Cycle 22 on this basis was 157.3

14.4.2 The initial growth rate parameters

On the other hand, R. M. Wilson (1990a,b) found a high correlation between the maximum value of the average growth rate of R_Z during the first years of the cycle and its value at the following maximum $R_Z(\text{max})$. He offered a prediction, based on the first 18 months of Cycle 22, of $R_Z(\text{max}) = 186.0 \pm 27.2$ and, based on the first 24 months, of 201.0 ± 20.1

Elling and Schwentek (1991) have introduced five parameters into the F-density function, known in statistical theory, and these parameters are varied to produce optimal fittings for each individual cycle and for the mean sunspot number curve (found by averaging cycles 10–21). Their forecasting technique involves making estimates of the values of these parameters, based on sunspot numbers recorded at quarterly intervals after the new cycle has begun. Their claim is that accurate forecasts can be made after 12 quarters (three years after sunspot minimum), and their 'best' prediction, based on data obtained over $3\frac{1}{2}$ years into Cycle 22, was $R_Z(\text{max}) = 174.7 \pm 11.9$.

While for some purposes forecasts with such limited lead time may be of value, almost any educated guess made during the period 1987–89 would have forecast that Cycle 22 would be bigger than average, probably bigger than Cycle 21, but not as big as Cycle 19. To be of value, forecasts based on post-minimum data might, reasonably, be required to produce a higher degree of accuracy than those based only on data available prior to minimum.

14.4.3 The shape and structure of the corona

At sunspot minimum, the corona near the poles consists of short, almost radial polar streamers extending into the region of low coronal density, the polar coronal hole. Near the equator, long equatorial streamers extend for several radii.

As the strength of the polar field increases, so does the angle at which the polar streamers are bent towards the equator (the polar-field bending angle PFBA). The latitude coverage of the coronal holes (CHS) increases, as does the extent of the equatorial streamers relative to the polar streamers.

Several studies have made use of these properties as proxies of the polar fields. Schatten *et al.* (1978) measured the PFBA, as did Schatten and Sofia (1987), while Schatten (1988, private communication) has investigated the size of the coronal holes.

Observing the corona is, however, not a simple task. Coronagraph observations are not of great value, because the occulting disk which blocks out the photosphere also tends to obscure the coronal features to be measured. Although eclipse photographs extend back to 1878, they are of variable quality and effectively random phase with respect to the cycle, and of insufficient frequency to permit reliable interpolation. Layden *et al.* (1991) discussed at length the various parameters, concluding that, whether or not they correspond to the polar fields, the PFBA is almost uncorrelated with $R_Z(\text{max})$, while the CHS and other structure parameters are, at best, weakly correlated with $R_Z(\text{max})$.

14.4.4 The polar faculae

Since photospheric faculae are known to be associated with the less concentrated magnetic field elements adjacent to sunspots, it would seem reasonable to suggest that counts of polar faculae near sunspot minimum might provide a simple physical proxy for the polar fields. Sheeley (1966) has, in fact, established a direct relationship between the number of faculae in the vicinity of the poles and the total polar magnetic flux.

Using Sheeley's data for each year between 1906 and 1976, Layden *et al.* have developed the equation for the regression line (cf. Equation 14.1)

$$R_Z(\text{max}) = 3.67(\pm 2.76)\{N_{\text{fac}} - 34.39(\pm 7.63)\}. \qquad (14.2)$$

Although the overall trend is encouraging, statistical tests show that the fit is not significant; in particular, the facular count corresponding to Cycle 16 is clearly anomalous.

Layden *et al.* also note that the maximum facular count and the minimum of R_Z frequently occur at different times. Since Sheeley's data do not extend past the minimum prior to Cycle 21, the method cannot be applied to Cycle 22. Layden *et al.* conclude that, at present, facular counts are not a useful precursor predictor of $R_Z(\text{max})$.

14.4.5 The structure of the interplanetary magnetic field

A neutral 'current sheet' close to the plane of the solar equator separates the interplanetary extension of the Sun's poloidal magnetic field (IMF). This sheet is warped in latitude away from the equator by weak equatorial 'sector' fields, so that there are, typically, four sectors of alternating polarity arranged in longitude around the Sun, which give the current sheet two northward and two southward facing maxima extending out through the inner solar system. At sunspot minimum, the warping typically has an amplitude of 15°, which varies from one minimum to the next, presumably in response to the intensity of the polar fields, the magnetic pressure of the stronger fields decreasing the amplitude of the warping angle. Schatten *et al.* (1978) and Schatten and Sofia (1987) have used the flattening of the current sheet at sunspot minimum to estimate the strength of the polar fields.

Unfortunately, satellite data do not cover a sufficient time-span, and, although earthbound estimates cover seven cycle minima, there are many practical difficulties associated with these measurements, not least of which is the influence of the Earth's magnetic field. After carefully reviewing these measurements, Layden *et al.* questioned whether the current sheet was sufficiently uniform to allow the measurement of a consistent flattening parameter by any technique with the accuracy needed to predict the activity maximum of the next cycle, a result confirmed by plots of $R_Z(\text{max})$ against the amplitude of the IMF.

14.4.6 *The geomagnetic indices* aa *and* Ap

The *aa* and *Ap* indices measure the response of the geomagnetic field to fluctuations in the solar wind. The *aa*-index is the average of the disturbances measured at two antipodal stations, while the *Ap*-index is the average of the disturbances measured at many stations distributed over the surface of the Earth; thus *Ap* is a better measure of global disturbances of the geomagnetic field. These geomagnetic disturbances arise in response to activity-modulated fluctuations in the solar wind. Data for *aa* are available from the National Geophysical Data Center spanning the years 1868–1989, and 1932–1989 for *Ap*. R. M. Wilson (1990b) has also used the average of the *aa*-index over the 36-month interval prior to sunspot minimum, *aa*(36).

Svalgaard (1977) extended the *aa* data for 1868–1976 with Wolf's calibration of daily magnetic declination variations for the period 1781–1880, supplementing the gap between 1805 and 1813 with counts of auroral displays so that, although his data are less homogeneous, the index, $aa^{(S)}$, significantly extends the data base. All these indices exhibit oscillations which are approximately in phase with the cycle, the minima of the indices occurring within about a year of the sunspot minima. It would seem that the combined influence of the strong polar and weak equatorial fields at this time is less than the influence of the strong equatorial fields near maximum.

It has been argued that the minima of the indices reflect the strength of the polar fields at sunspot minimum. However, the contributions of the equatorial fields at this time, although small, cannot be neglected. Thus, although the value of these indices as *proxies* for the polar fields must be regarded as somewhat doubtful, as *precursors*, they have proved to be the most reliable indicators of $R_Z(\max)$. Linear regression analyses applied to the data pairs $(aa_{\min}, R_Z(\max))$ yield the statistically significant relations

$$R_Z(\max) = 8.59(\pm 1.37)\{aa_{\min} + 1.13(\pm 0.67)\} \tag{14.3}$$

and

$$R_Z(\max) = 20.38(\pm 4.88)\{Ap_{\min} - 2.06(\pm 0.35)\}. \tag{14.4}$$

These relations gave values of $R_Z(\max)$ for Cycle 22 of $160(\pm 28)$ and $162(\pm 43)$. A similar analysis applied to Svalgaard's data $aa_{\min}^S$ yields the regression relation

$$R_Z(\max) = 11.1(\pm 1.44)\{aa_{\min}^S - 2.38(\pm 0.50)\}, \tag{14.5}$$

which gives $R_Z(\max) = 182(\pm 26)$ for Cycle 22.

In an independent analysis, R. M. Wilson (1990b) distinguishes between annual averages and moving averages for the geomagnetic indices. For aa he obtained essentially the same relation as Equation (14.3) for either average, with Pearson correlation coefficients of 0.89 and standard errors of 21.5. For $aa(36)$ he obtained the relation

$$R_Z(M) = 5.90aa(36) + 10.57, \tag{14.6}$$

with correlation coefficients and standard errors of 0.90 and 21.0, while for Ap_{ma} (moving average) he found

$$R_Z(M) = 18.56Ap_{\mathrm{ma}} - 24.23. \tag{14.7}$$

Wilson's predictions for Cycle 22 were 171.3 and 169.7, based on annually averaged values of the aa and Ap indices, 161.3 and 161.4 based on moving averages, and 160.9 based on $aa(36)$.

Thompson (1988) has analysed the structure of the geomagnetic disturbance profile through the cycle and his results (see Figure 14.1) compare the number of days per year on which the Ap-index equalled or exceeded 25 with the smoothed sunspot number. This result clearly shows that the Ap-index exhibits two peaks during any given cycle, one approximately coincident with sunspot maximum, and the other late in the declining phase. Thompson has identified the first with particle streams, which are generally thought to be associated with solar flares (but see § 10.9 above and § 14.7 below), and showed that there is some tendency for the amplitude of this peak to correlate with the current cycle. He noted, however, the very peculiar case of 1980, when, despite a normal number of flares, the geomagnetic disturbances were abnormally low.

The amplitude of the second peak does not correlate well with $R_Z(\max)$ for

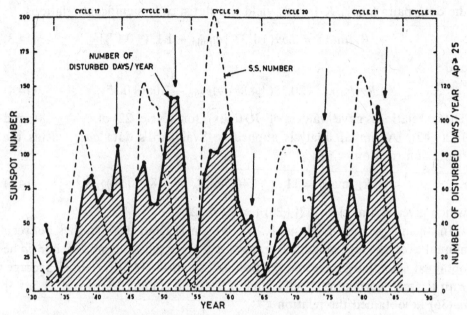

Fig. 14.1 The annually-averaged sunspot number since 1930 is represented by the dashed line, and the annual number of geomagnetically disturbed days since 1932 is represented by the solid line. (From Thompson 1988)

the current cycle, but there does appear to be a correlation with $R_Z(\text{max})$ for the next cycle. Using this correlation, Thompson (1985) forecast a sunspot maximum of 159 for Cycle 22.

Other authors, using geomagnetic data, made predictions for $R_Z(\text{max})$ of 165 ± 35 (Kane, 1989) using *aa* data, and 160 ± 40 (*aa*) and $165 \pm 35(Ap)$ (Gonzales and Schatten 1987).

14.4.7 *Bi-variate analyses*

R. M. Wilson has considered several combinations of precursors in order to obtain improved fits between precursors and $R_Z(\text{max})$. His most successful combination is between annual averages $Ap(\text{min})$ and $R_Z(\text{min})$, which takes the form

$$R_Z(\text{max}) = 27.78 - 5.86R_Z(\text{min}) + 17.75Ap(\text{min}), \qquad (14.8)$$

for which the Pearson coefficient, based on cycles 17–21, is 0.997, and the standard error is 3.9. Although these correlation parameters are impressive, it should be noted that this relation now involves a three-parameter fit, and the extra degree of freedom naturally makes it possible to improve what for the *aa* indices were good two-parameter fits. The prediction for Cycle 22 was 144.6($\pm$2.5).

Thompson has more recently (1992) proposed a bivariate analysis of a different kind. Using his finding that the magnetic disturbance distribution exhibits two maxima during a given cycle, the first correlated with R_Z(max) for the current cycle, R_c, and the second with R_Z(max) for the next cycle, R_n, he has applied a multiple linear regression analysis to his data, yielding the relation

$$N_m = -47.9 + 2.20R_c + 1.88R_n, \tag{14.9}$$

where N_m is the total number of magnetic disturbances during the cycle. The amplitude of the next cycle may thus be predicted once R_c and N_m are known. Application to past cycles yields an rms error of 17 and a forecast for Cycle 22 of 148.3

14.5 Discussion–the sporting journalist's report

Since the bottom line in assessing any forecasting system is not its compatibility with existing results but its success in predicting a new result, it is now time to compare the forecasts of R_Z(max) for Cycle 22 with the 'declared official result' of 158.5. Perhaps the writer can be forgiven for borrowing from the sporting journalists, but it would seem that, by and large, astrophysicists' attempts to forecast the maximum amplitude of Cycle 22 were not much better than those of an average punter during a difficult day at the races.

The cycle got off to a great start and, by the first bend, had established a commanding lead (many predicted that it would shatter the existing record). Approaching the home turn, however, it faltered and dropped back, leaving many punters with empty wallets. Figure 14.2, adapted from R. M. Wilson (1990b), shows how the major competitors in the 'Predictors Stakes' went to the line.

This event was won by Thompson's estimate of 159, based on the second peak in the number of geomagnetic disturbances. Second, by a 'short half head' was the rank outsider with the worst track (correlation) record, R. M. Wilson's forecast of 157.3, based on R_Z(min). Not far away were some of the better fancied entrants from the geophysical indices stable, the Layden *et al.* prediction for aa_{min} of 160 ± 28 and R. M. Wilson's values based on (i) *aa*(33) of 160.9, (ii) aa_{min}, 161.3 and Ap_{min} (with moving averages), of 161.4. Unfortunately for this stable, Layden *et al.*'s prediction based on $aa_{min}^{(S)}$ finished further away at 186 ± 26. The best result for the connections of numerological methods was P. A. Smith's prediction of 160.

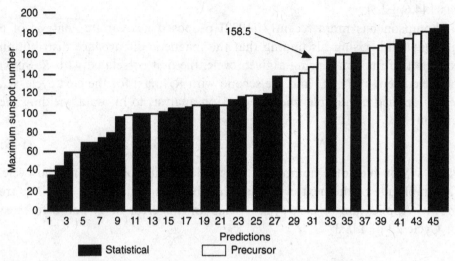

Fig. 14.2 The histogram compares a selection of the predictions for the maximum of Cycle 22 with the actual value of 158.5. The predictors are identified below. (From Thompson 1992, adapted from G. Brown)

1. Gregg, 1984
2. Cohen & Lintz, 1974
3. Oksman, 1984
4. Landscheidt, 1984
5. Cole, 1973
6. Otaola & Zenteno, 1983
7. Wilson, 1988a
8. Hedeman & Dodson-Prince, 1984
9. Berger, Goossens & Pestiaux, 1984
10. Xu, Zhao, Mei & Guo, 1979
11. Hill, 1977
12. Holland & Vaughan, 1984
13. Wang, Tang, & Zhang, 1984
14. Chistyakov, 1982
15. Tritakis, 1984
16. Kontor et al., 1983
17. Wu, 1984
18. Schatten & Hedin, 1984
19. Wilson et al., 1984
20. Kopecky, 1986
21. Lantos & Simon, 1987
22. Suda, 1980
23. Kapoor & Wu, 1982

24. Sargent, 1987
25. Schove, 1955
26. Wu, 1984
27. Hunter, 1980
28. Kataja, 1984
29. Wilson, 1990a
30. Wilson, 1988b
31. F. M. Smith, 1976
32. Thompson, 1985
33. P. A. Smith, 1984
34. Gonzalez & Schatten, 1987
35. Thompson, 1987
36. Kurths & Ruzmaikin, 1990
37. Wilson, 1990a
38. Schatten & Sofia, 1987
39. Fox, 1990
40. Brown, 1988
41. Lindberg, 1989
42. Wilson, 1990b
43. Kane, 1987
44. Butcher, 1990
45. Wilson, 1990b
46. Koons & Gomey, 1990

Among the more fancied runners who disappointed were the members of the bivariate team. Despite their impressive track records (correlation coefficients ≥ 0.96 and small standard errors), each finished at least one, and in some cases two or three standard errors off the correct result.

One might also have expected that predictors who placed their bets after the start of the event (a trick which racegoers have from time to time attempted, generally without success) would have achieved better results. Forecasts made after sunspot minimum, however, fared even worse than the bivariate forecasts: R. M. Wilson's estimate after 24 months (201) was less accurate than that after 18 months (186). Even with data extending $3\frac{1}{2}$ years into the cycle, Elling and Schwentek's value of 174.7 was further than their standard error of ± 11.9 from the correct result. Most disappointing of all was the newcomer to the field, the directly measured polar field, which gave values in the range 180–215.

14.6 Discussion–with hindsight

Briefly, the vital statistics for Cycle 22 are as follows:

(i) The cycle began in September 1986, from the highest minimum so far recorded, $R_Z(\text{min}) = 12.3$.

(ii) The early values were the highest on record.

(iii) The cycle achieved $R_Z(\text{max}) = 158.5$ in July 1989.

(iv) It is third ranked after Cycles 19 ($R_Z(\text{max}) = 201.3$) and 21 ($R_Z(\text{max}) = 164.5$).

(v) It featured the shortest rise time from minimum to maximum (35 months).

Despite the somewhat disappointing results for physical modelling methods, there are some hopeful indications. From the strict forecasting standpoint, it is clear that single-variate methods based on the geophysical precursors have come through well. Although the $aa-Ap$ single variate correlations with $R_Z(\text{max})$ for past cycles are not as impressive as those achieved by bivariate methods, the latter failed the test of a genuine forecast. With its extra degree of freedom, a new regression analysis could, no doubt, provide a modification of Equation (14.8) to yield a better correlation coefficient for all cycles, including Cycle 22. It is elementary, however, that three-parameter models which must be modified after the event are of less value than two-parameter models which require no further modifications.

For forecasting based on the strength of the polar fields at minimum, however, the result is disappointing. Although Layden *et al.* argue that the aa_{min} values are proxies for $B_P(max)$, these values must also reflect an unknown contribution from the equatorial fields. Since the more successful bivariate models involve a negative contribution from $R_Z(min)$, it could be argued that this subtracts out the contribution to the geomagnetic index from the equatorial fields. However, the relative failure of the bivariate models for Cycle 22 indicates that the various interactions among these quantities are not simple. In particular $R_Z(min)$ is a composite sum of the tail of the old cycle and the beginning of the new. If $R_Z(min)$ is large, as it was before Cycle 22, this does not necessarily imply that the amplitudes of either cycle are relatively large, rather it may be that they share a greater degree of overlap.

Proxies which have been proposed for the polar fields must remain uncertain until we have more evidence that they correlate with actual measurements of polar fields. By then we will be able to explore the correlation between $B_P(min)$ and $R_Z(max)$ directly, and proxies will be of value only in order to extend the data base for $B_P(min)$, if the direct correlation appears successful. Unfortunately the results for Cycle 22 do not augur well for this prospect.

14.7 The role of coronal holes

In his study of the geomagnetic disturbance profile, Thompson (1988) has suggested that, while the first peak is associated with the current cycle, the first effect of the new cycle is to 'shut down' the geomagnetic disturbances caused by the effects of the old cycle, as represented by this peak. Thompson draws particular attention to the early arrival of the second peak of geomagnetic activity during Cycle 21 (see Figure 14.1) and notes that this was preceded by a period of unusually low geomagnetic activity, centred on the year 1980. This was a year of high solar activity, close to the maximum of the second largest cycle on record. Yet, despite the occurrence of the expected number of solar flares, the level of geomagnetic activity was abnormally low. Thompson suggests that this might be connected with the early arrival of precursor geomagnetic activity in 1982, postulating that the first effect of a new cycle is not the enhancement of geomagnetic activity in the declining phase of the old cycle, but rather the *suppression* of the geomagnetic activity associated with the old cycle, prior to the onset of the precursor activity associated with the new cycle.

Geomagnetic activity around solar maximum is usually attributed to the particle emissions related to solar flares and associated filament disappearances. Thompson's suggestion that geomagnetic activity can be inhibited by the early effects of the new cycle implies that the flares which occurred in 1980 were made, in some way, less 'geo-effective'. Normally, X-class flares are followed, after two days, by significant geomagnetic disturbances, but, of the 18 X-class flares occurring in 1980, not one was followed by a day of geomagnetic disturbance.

While 1980 may represent an extreme example of the suppression of geomagnetic activity prior to the onset of precursor disturbances, Thompson notes other possible cases. Geomagnetic activity during Cycle 18 showed a pronounced dip just after sunspot maximum, and this preceded the strong geomagnetic activity which was a precursor to the record Cycle 19. Dips in geomagnetic activity also occurred during the declining phases of Cycles 17 and 20, but Cycle 19 illustrates the opposite situation in which geomagnetic activity tracked the sunspot cycle quite well; precursor activity was almost negligible and was followed by the weaker Cycle 20.

Here we have the interesting situation where, by seeking to base forecasts on precursors to the cycle, Thompson has obtained correlations which challenge and may inform our physical understanding of the cyclic processes. Hewish (1988) has argued that all geomagnetic disturbances arise from coronal holes, rather than from solar flares, and, if this is the case, the absence of geomagnetic disturbances following the flares of 1980 may be understood in terms of a significant change in the number and location of the coronal holes in 1980, as the *first* precursor to Cycle 22.

At the very least, Thompson's research supports the arguments of § 10.9 that the surface distribution of coronal holes during the cycle plays an important role in the cyclic processes. This role is, at present, imperfectly understood, and a major task of future research and forecasting efforts should be the study of the sub-surface structure of the coronal hole and its relation to flaring active regions.

14.8 Forecasting for Cycle 23 and beyond

Forecasters now face a difficult choice. If the solar cycle is a genuinely chaotic system, in which small changes in the initial conditions can produce widely divergent trajectories, then our attempts to make long-term forecasts of the characteristics of future cycles are doomed to failure. No matter how carefully we record the parameters of the previous cycle or the early patterns of the new one, we cannot hope to predict either the time or the amplitude of the maximum with any confidence. The failure of forecasts which attempt to determine long-term periodicities in the records of previous cycles suggests that the system may, indeed, be chaotic. The rapid rise of Cycle 22 led some

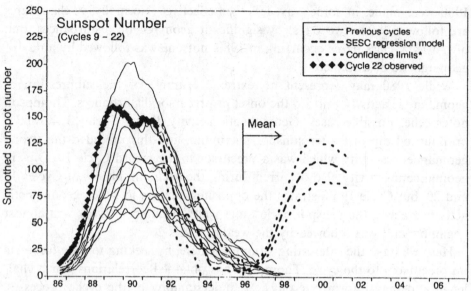

Fig. 14.3 A forecast of the declining phase of Cycle 22 and the rising phase of Cycle 23 produced by NOAA using a regression analysis.

to forecast that this would be the strongest cycle recorded. However, its sudden loss of momentum in late 1988 and premature peak in June 1989 is very suggestive of chaotic behaviour and left forecasters in a state of some embarrassment.

Nevertheless, just as long-range weather forecasting is still actively pursued (perhaps on the grounds that planners would prefer to have *some* forecasts on which to base decisions rather than none at all), the only forecast that can be made with complete confidence is that solar physicists will continue to make forecasts of future cycles. Indeed, an early forecast of the declining phase of Cycle 22 and the rising phase of Cycle 23, produced by NOAA and based on a regression analysis, has already been published and is shown in Figure 14.3. Strategies by which such forecasts might be improved are now considered.

Although we have dismissed the simple form of the Babcock model as the basis for a physical model, the concept of the *extended cycle*, discussed in Chapter 8, deserves further study, since it entails that some features of the new cycle appear at mid-latitudes early in the previous cycle. At or near maximum of that cycle the synoptic magnetic pattern bifurcates, one broad branch moving rapidly polewards and the other moving more slowly towards the equator. The poleward-moving branch is broad, and the arrival of its centre of gravity at the polar region at sunspot minimum corresponds to the maximum of the polar fields. The equatorward branch progresses more slowly and, of course, coincides with the sunspot activity at lower latitudes.

If these branches are genuine products of a bifurcation of a single physical system, then one would expect a strong correlation between $B_P(\text{min})$ and the following $R_Z(\text{max})$, for reasons which do not involve acceptance of the details of the Babcock model. Further, the mid-latitude magnetic field signal at and prior to the bifurcation (occurring near the maximum of the previous cycle) should provide a physical precursor for the following cycle, permitting forecasts of $R_Z(\text{max})$ at least half a cycle earlier than those based on $B_P(\text{min})$ and having the same or greater physical justification.

How the new cycle manages to shut down the geomagnetic effects of the old cycle, as Thompson suggests, and what role is played by the coronal hole patterns, are questions for which there are, at present, no adequate answers. However, data regarding this sudden drop in the geomagnetic indices do appear to offer the possibility of early indications of the likely strength of the next cycle.

It will, of course, require some time before an adequate data-base concerning the global poloidal fields used to construct Figure 8.7 can be gathered, but, if appropriate correlations can be established, this greater lead time would provide forecasters with a significant advantage.

References

Brown, G. M.: 1988, *Nature*, **333**, 121.

Elling, W., and Schwentek, H.: 1991, *Solar Phys.*, **137**, 155.

Gonzalez, G., and Schatten, K. H.: 1987, *Solar Phys.*, **114**, 189–92.

Hewish, A.: 1988, *Solar Phys.*, **166**, 195.

Kane, R. P.: 1989, *Solar Phys.*, **122**, 175.

Layden, A. C., Fox, P. A., Howard, J. M., Sarajendini, A., Schatten, K. H., and Sofia, S.: 1991, *Solar Phys.*, **132**, 1–40.

Schatten, K. H., Scherrer, P. H., Svalgaard, L., and Wilcox, J. M.: 1978, *Geophys. Res. Letters*, **5**, 441.

Schatten, K. H., and Hedin, A. E.: 1984, *Geophys. Res. Letters*, **1**, 873.

Schatten, K. H., and Sofia, S.: 1987, *Geophys. Res. Letters*, **14**, 632.

Schatten, K. H.: 1988, *private communication*.

Sheeley, N. R. Jr.: 1966, *Astrophys. J.*, **144**, 728.

Svalgaard, L.: 1977, in J. Zirker (ed.), *Coronal Holes and High Speed Wind Streams*, Colorado Associated University Press, Boulder, Colorado.

Thompson, R. J.: 1985, *IPS Technical Report IPS-TR-85-06*, IPS Radio and Space Services, Sydney, Australia.

Thompson, R. J.: 1988, *IPS Technical Report IPS-TR-88-01*, IPS Radio and Space Services, Sydney, Australia.

15

Summary and conclusions

...the King said, very gravely, 'and go on till you come to the end: then stop.'

Lewis Carroll

15.1 Introduction

It is now time to summarize and conclude but, regrettably, it is not possible to follow the King's advice, for the end of this story is not yet in sight. As in any good detective story, both clues and red herrings have been scattered liberally throughout the foregoing chapters. Unlike the novelist, however, we cannot be certain which is which and so cannot now discard the red herrings, marshal the significant clues, and reveal the solution. At best, we can indicate what appear to be the important themes which must form part of the solution and suggest the crucial areas of investigation which may lead to a satisfactory resolution.

Two important questions currently remain without clear and unequivocal answers:

(i) Do both cyclic and non-cyclic stellar activity arise as a result of dynamo action and, if so, of what kind (or kinds)?

(ii) Is the aperiodicity of solar cyclic activity evidence of a chaotic system, and, if so, can the existence of a strange attractor be demonstrated and its dimension determined?

Until firm answers can be provided for these questions, the problem of cyclic stellar activity remains without a satisfactory solution.

261

15.2 The input from stellar cycles

Because of the more extensive data-base regarding the solar cycle, the discussions in this volume are weighted heavily in that direction, but this in no way implies a greater importance to solar over stellar studies. Although still in their infancy, studies of stellar activity have already made a significant contribution to our understanding of the problem of cyclic activity.

Activity in young, rapidly rotating stars is more violent but less regular than in older stars, such as the Sun. As stars age, however, both their rotation rates and their activity decline, and some exhibit cyclic properties with periods similar to that of the solar cycle. In addition to rotation, the depth of the convection zone appears to be crucial to the activity patterns. F-type stars, with their shallow convection zones, exhibit sporadic activity patterns. Older G-type stars, such as the Sun, develop more extensive convection zones which appear to be a necessary, but not a sufficient, condition for cyclic activity. Fully convective late M-type stars exhibit strong but irregular activity.

In Chapter 7, it was concluded that cyclic activity depends on a balance between the extent of the convection zone and the rate of rotation. The observations reported there show that an 'adequate' (i.e. not too shallow) convection zone is necessary, as is an appropriate rate of rotation. When the convection zone is too shallow, activity is present but cyclic patterns are not detected; when the convection zone is too extensive, cyclic activity is overcome by stronger but less regular activity. Again, for rotation periods of less than 20 days, activity tends to be strong but irregular; for longer rotation periods, activity tends to be less intense and cyclic, with periods ranging from eight to fifteen years. The cyclic activity exhibited by the Sun is consistent with this pattern, and it would seem likely that the older, more slowly rotating, cyclically active stars share a common mechanism for creating and modulating their activity.

15.3 Dynamo models of cyclic activity

Accepting this as a working hypothesis, we must seek the mechanism responsible for solar-type activity cycles. Many, but not all, workers regard the involvement of some form of dynamo action in stellar activity and activity cycles as almost self-evident. In Chapter 11 it was shown that, according to the mean field formulation, the α–ω dynamo depends on the interplay between gradients in the velocity fields associated with rotation and with cyclonic convection in the presence of a magnetic field, the former providing the

ω-effect and the latter the α-effect. It is in this context that results concerning stellar cycles obtained over the past decade are particularly relevant.

This balance between the effects of convection and rotation is characterized by the Rossby number, the ratio of the rotation period to the turnover time of the largest convection eddies. Young, rapidly rotating stars with extensive convection zones have a low Rossby number and exhibit strong but sporadic activity (as indicated by the strength of the CaII emission). However, *cyclic* activity appears only in older stars having significant, but not too extensive, convection zones and rotation periods exceeding 20 days, i.e. they have a larger Rossby number.

In order to understand this, it may be helpful to review Parker's simple picture (§6.8) of an Ω-loop, which is formed from a subsurface toroidal field and twists as it rises under the influence of the Coriolis force. In this context it is postulated that, in rapidly rotating stars, the α-effect produced by the Coriolis force is dominant. The Ω-loop is twisted randomly and loses all sense of the orientation of the subsurface toroid, so that the resulting activity is strong, but irregular.

On the other hand, if the loop is twisted through an angle of order 90°, and less than 180°, the orientation of the regenerated poloidal field retains some relation to that of the toroidal field, so that the dynamo wave can propagate in latitude with some degree of regularity and in the appropriate sense according to Parker's heuristic picture. This picture is undoubtedly an oversimplification, but the general result that cyclic activity flourishes when the α-effect produced by the Coriolis force is in some form of balance with the rotational gradients which produce the ω-effect is consistent with the α–ω dynamo account of solar-type cyclic activity.

The α–ω mode of dynamo action is, however, not the only type of dynamo which might operate in a star. In a rapidly rotating star the radial gradients may not be large, and the α-effect may dominate the ω-effect. In this situation, the second term on the right side of Equation (11.16) may be comparable with or greater than the first, and dynamo action may occur in the α^2–ω, or α^2 modes, rather than in the α–ω mode. Gray has suggested (see §11.3) that, as luminosity class III giant stars evolve across the 'granulation boundary', they have at best only shallow convection zones within which the radial gradients would be small, and that the activity which appears in these stars at this phase of their evolution is generated by an α^2 dynamo. An α–ω dynamo is generally thought to operate in a layer or shell below the star's convection zone, but, in fully convective late M-type stars, there is no possibility of a shell dynamo. Thus the strong but disordered activity exhibited by these stars may be ascribed to a fibril dynamo, acting in some region of the convective star and operating in the α^2 mode (Weiss 1993).

These results suggest that, over a broad spectrum of stellar parameters, the occurrence of activity is consistent with the MFE formulation, which shows how α–ω, α^2, and even α^2–ω dynamos might occur under different circumstances (see Equations 11.16 and 11.17). It would appear that some form of dynamo action is responsible for generating stellar activity and that different modes may be involved in the transition from irregular activity in younger stars to more regular cyclic activity in older ones. Although neither the MFE kinematic models, nor the non-linear dynamical formulations, nor the fast-dynamo models have yet provided a fully satisfactory account of solar and stellar activity, the problem would seem to lie in our inability to formulate the mathematical problem succinctly and accurately rather than in the physical concepts of the astrophysical dynamo.

Critics of this view, such as D. O. Gough (private communication), stress that, while the mean-field equations derived in § 11.3 may be applicable in some astrophysical circumstances, they are obviously not valid for the Sun, where the perturbing field may exceed the mean field by several orders of magnitude. Gough is also concerned that, while the periods of cyclically active stars (ranging from 8 to 15 years) do not arise naturally from mean-field theory, they can be obtained from estimates of the frequencies of torsional oscillator models (see § 6.7). Piddington (1972) has also stressed the problems created for dynamo models by the high conductivity, and therefore low diffusivity, of stellar plasmas. Some possible answers were canvassed in § 11.5, but the matter remains to be fully resolved.

Although these may be valid criticisms, the critics become less confident when asked to put forward any workable alternative models. Oscillator models retain their adherents, and, as noted in § 12.11, some preliminary synoptic data from helioseismology are not inconsistent with an adiabatic exchange of energy between rotation and the magnetic field. These results are tentative and cannot yet be regarded as providing firm support for torsional oscillator models, but they offer the prospect that, within the next decade, helioseismology may provide a resolution to this question. To date, however, there has been no satisfactory answer to the objections to oscillator models which were discussed in § 6.7.

It seems that, despite its drawbacks, dynamo action in one or more of its several modes will continue to be widely used to account for the various forms of stellar activity because, like democracy, it is preferable to any of the available alternative systems.

15.4 Chaos and cyclic activity

While it is of value in providing insight into some aspects of astrophysical dynamos, the kinematic (i.e. linear) dynamo cannot provide a fully satisfactory model to account for activity in the Sun and solar-type stars because it neglects the effects of the Lorentz forces which, eventually, must influence the velocity fields and limit the growth of the magnetic fields. As Equation (11.23) demonstrates, the dynamical equations are non-linear in the magnetic field, and studies of such systems (summarized in Chapter 13) show their susceptibility to chaotic behaviour.

For the Sun, some elements of chaos are indicated by the *aperiodic* form of the sunspot number variations. No two cyclic profiles are the same, but each one exhibits similar features: the ~ 11-year periods; the rapid rise to maximum; the rather irregular form of the maximum, during which the polar fields reverse; the sharp initial decline of sunspot activity after the reversal; and the more gentle decline to minimum. There is a parallel between this behaviour and that of a strange attractor, such as those shown in Figure 13.3. Further, a plausible interpretation of the occurrence of grand minima, such as the Maunder Minimum, is that they correspond to a jump to another branch of the attractor.

Stellar studies have shown that, by comparison with the Sun, some solar-type stars exhibit low levels of activity in which it is difficult to detect cyclic patterns. Baliunas and Jastrow have argued (see § 7.11) that these stars are indeed 'normal' solar-type stars which are passing through grand minima, similar to the Maunder Minimum. If this interpretation is correct, it would seem that activity in solar-type stars generally exhibits elements of chaotic behaviour.

To demonstrate chaotic behaviour unambiguously, it is necessary to determine the dimension of the strange attractor. Unfortunately, as § 13.7 shows, this would require a data base several orders of magnitude larger than the sunspot number records since 1640. The presumption that solar and stellar activity cycles are manifestations of nonlinear chaotic systems is, however, compelling and has important implications for forecasting the parameters of the solar cycle (see § 15.6).

15.5 Significant features of the solar cycle

The study of the range of modes of stellar activity across the spectrum of stellar parameters provides information concerning the different modes of dynamo action which may apply across this spectrum. Recent solar results provide information concerning details of the mechanisms involved in the solar dynamo.

15.5.1 *The extended activity cycle of the Sun*

The extended activity cycle (EAC), discussed in Chapter 8, raises the fundamental question as to whether each cycle can be thought of as a unique entity, with an identifiable beginning and an end, and, if so, whether it is appropriate to define each cycle to begin at sunspot minimum and to end at the following minimum, or whether phenomena which can be identified with different cycles can co-exist on the Sun and, if so, for how long. Of course, no cycle can be completely divorced from its predecessor, but, in attempting to understand the important physical processes, it is not unreasonable to look for those phenomena which, when identified, can best characterize the cycle which follows.

The obvious overlap of the butterfly diagram leaves no doubt that adjacent cycles do overlap, and that the determination of the time of sunspot minimum represents a balance between the decay of the old cycle and the rise of the new. The discussions of Chapters 8, 10, and 14 sought to probe the extent of the overlap, partly for the purpose of forecasting but, more fundamentally, to explore the nature of the physical processes involved.

In the Babcock model, the cycle is assumed to begin at, or just prior to, sunspot minimum and the model takes, as its starting point, an assumed global poloidal field from which the toroidal fields are generated. Some of the difficulties for relaxation models of this type were explored in Chapter 6, but if, as argued in Chapter 8, the first phenomena of the new cycle emerge at or before the minimum of the previous cycle, such models are untenable.

Dynamo wave models, such as that of Parker (§ 6.8), are consistent with the co-existence of two migrating cycles in each hemisphere, although, as discussed in Chapters 6, 11, and 12, they encounter problems regarding the fundamental direction of propagation indicated by the butterfly diagram. The most recent EAC data, such as that illustrated in Figure 8.7, in which the new cycle begins at mid-latitudes at or before the maximum of the old cycle, and bifurcates into a poleward and an equatorward branch, place greater constraints on such models.

These provide a challenge for the more sophisticated generation of astrophysical dynamos discussed in Chapter 11 and above. Results concerning the evolution of the large-scale fields and the polar fields reversals are also relevant to this problem.

15.5.2 Small- and large-scale fields

An important conclusion from Chapters 9 and 10 is that the large-scale fields, which play a significant part in the cyclic processes, arise not just from the decay of active-region fields but from flux emergence at all scales of the bipole size spectrum. This is an important result which is gradually gaining recognition. Although the largest emerging bipoles are probably generated deep within the convection zone or below it, the small-scale fibril fields, which continually emerge and renew the large-scale patterns, may arise from dynamo action occurring at any depth (including the upper layers of the convection zone) and at any latitude at which the appropriate velocity gradients and cyclonic motions are found.

Although the lifetimes of these small-scale fibril fields are only a few days, the large-scale patterns to which they contribute may last for several years. The equatorial pattern (or streak) shown in Figure 9.9, for example, extends from 1984 to 1988, and the longevity of the large-scale patterns argues for a coherent long-term source of the emerging fibril fields. The inclination of the streaks in the stackplots indicates the rotation rate of this source, which probably corresponds to the rotation rate of the level at which the source is located, and the observed changes in the inclination of the streaks in the stackplots may indicate changes in the rotation rate at that level or a change in the level of the source which continues to replenish the surface fields. Intersecting streaks seen in the patterns suggest that some surface features at the same latitude may arise from sources at different depths.

The longevity and the multiple rotation rates exhibited by these streaks suggest that their source may be related to large-scale, sub-surface velocity fields or cells which continuously generate and expel flux towards the surface, rather than to the passive diffusive decay of active region fields. Further, the asymptotic decrease in the gradients of the streaks just prior to their termination, which is observed, for example, near sunspot maximum and the reversal of the polar fields (see Figure 9.3(b)), suggests the occurrence of a fundamental change in the configuration of these cells.

Legrand and Simon (1991) have argued for the existence of a two-component cycle in which the polar dipole field is generated not by the decay and poleward drift of active-region fields *following* the sunspot cycle but by an independent (albeit unspecified) process which precedes but is linked to the *next* sunspot cycle. The possibility that this process is related to a change in the large-scale velocity patterns is attractive and may be resolved by future synoptic results from helioseismology. This suggestion is in direct conflict with the Babcock–Leighton–Sheeley flux-transport model; nevertheless, the arguments put forward by Legrand and Simon, by Stenflo, and by the author and his colleagues (reviewed in Chapter 10), together with the discussion of the extended cycle (Chapter 8), suggest that an understanding of cyclic phenomena lies in this direction rather than in further attempts to fine-tune the flux-transport model.

Although Legrand and Simon do not specify the mechanism by which the sub-surface dipole field is generated, we speculate that it may be related to the migration of the polar crowns from latitudes $\sim 50°$ towards the poles, where their arrival coincides with the reversals and the build-up of the new polar fields. McIntosh (see §9.4 and §10.2) argues that this migration must reflect some fundamental sub-surface process rather than result from surface flux migration. The surface signatures of this process are the local poloidal fields which define the crown, and it may be that the toroidal component of an α–ω dynamo is confined below the surface at some level where the rotational gradients favour poleward propagation.

When the wave reaches the polar regions, the polar singularity becomes important and the simple analysis of §11.4 is no longer valid. Numerical solutions of the dynamo wave equations near the singularities of spherical or cylindrical polar coordinate systems are unstable, and observations are the best guide to subsequent developments. McIntosh has identified the dark diagonal streak in the north polar stackplot of Figure 9.3(b) (a white streak in the south) between 1966 and 1968 with the polar crown gaps, and the abrupt termination of these streaks in 1968 marks the closure of the polar crown gap which appears to set up the polar field reversal. As argued above, the decrease in the gradients of these diagonal streaks just before the patterns terminate may represent some more fundamental development in the internal rotational structure of the region at this time.

As these axisymmetric poloidal fields continue to grow during the declining phase of the old cycle, so the corresponding sub-surface toroid should grow and expand outwards, i.e. equatorwards. The sudden decline in old-cycle activity following shortly after the polar field reversals and the subsequent rise in precursor geomagnetic activity during the declining phase of the cycle point to the likelihood of a significant change in the arrangement of the sub-surface toroidal fields.

Although we cannot attempt to describe the nature of these changes on the basis of the rather sketchy data available, the numerical simulations of Brandenburg *et al.* (and others, see §11.6) may provide a clue, by showing how compressible convection in a stratified rotating system generates flux tubes which wind around sinking plumes with locally concentrated vorticity, thus generating elements of strong toroidal fields. The simulations of

Brandenburg *et al.* refer to a cartesian box geometry, but the singularity of a polar geometry may well increase the likelihood of the generation of the strong axisymmetric toroidal fields about the poles which are necessary to support the next sunspot cycle. It is, perhaps, now time to attempt solutions of the dynamo equations in polar coordinates for a region about the pole.

This is a formidable task and must involve some knowledge of the configuration of the large-scale velocity fields involved in these processes. Snodgrass and Wilson (1987) have postulated the presence of large, axisymmetric doughnut-shaped convective rolls near the poles, and, while improved frequency-splitting data may provide some guidance, it must be remembered that the interpretation of the splitting data in terms of the rotational structure of the solar interior is least reliable at high latitudes.

15.6 Precursors and forecasts

Finally we return to the problems of forecasting the parameters of the cycle. These forecasts are important not only on account of the need to predict future terrestrial conditions associated with the sunspot cycle but also because successful forecasts based on physical hypotheses demonstrate some improved understanding of the phenomena.

Although confirmation of the chaotic nature of the cyclic process would eliminate the possibility of accurate long-term (i.e. > 50 year) forecasting, this does not necessarily mean that we cannot understand the physical mechanisms and thus follow or predict the short-term trajectory of the cycle. There is a global pattern to cyclic phenomena, and our minds, conditioned by Newtonian training, tend to resist the notion that a system which displays the regularity of the solar cycle must lie beyond the reach of mechanistic explanation and prediction. In particular, short-term forecasts of, say, the magnitude of the $(n + 1)$th cycle at various epochs within the nth cycle have achieved some successes, and these encourage further endeavours to understand the relationship between the precursors and the 'main event'.

Of the various precursors discussed in Chapter 14, perhaps the most interesting and challenging is the possibility, suggested by Thompson (1988) (illustrated in Figure 14.1), that the geomagnetic effects of a given cycle are in some way 'switched off' shortly after maximum by the first phenomena of the new cycle: these subsequently give rise to a peak in the geomagnetic indices late in the old cycle, the amplitude of which correlates with the amplitude of R_Z in the next cycle. Although this connection is far from proven, Thompson's forecast of R_Z(max) for Cycle 22 was closest to the true value (see §14.5), and his suggestion that the geomagnetic effects are related to the occurrence of coronal holes in the neighbourhood of active regions, rather than to the flares emitted by these regions, raises interesting possibilities.

It is worth emphasizing that, whereas coronal holes are defined (see § 3.10) in terms of the above-surface configurations of their magnetic fields, there is at present no model for the sub-surface configuration of the fields associated with these holes, nor of their connection with the sub-surface fields which give rise both to strong activity complexes adjacent to the coronal holes and to the large-scale fields. McIntosh (see § 10.9) believes that the evolution of the equatorial coronal holes at mid-cycle is related to the growth of the polar holes and, using He 10830 stackplots (see Figure 10.7), has described some interesting examples of the evolution of equatorial holes during Cycle 21. These questions should figure prominently in future research, not only to improve cyclic forecasting but also to obtain a better understanding of the mechanisms of cyclic activity.

15.7 Conclusion

Studies of the evolution of solar-type stars have already provided information concerning the past history of the Sun and a look into its future. Investigations of other F, G, K, and M-type stars have indicated why some forms of stellar activity are cyclic and why others are not. Synoptic studies of the surface distribution of magnetic fields, particularly in relation to coronal holes, will establish more precise boundary conditions governing the three-dimensional structure and evolution of the global solar magnetic fields. Further studies of non-linear and fast dynamos may resolve some of the problems surrounding the application of dynamo theory to solar and stellar cyclic activity patterns. Progress in chaos theory holds out the prospect of a better understanding of aperiodic non-linear systems, and future developments in helio- and even stellar seismology may instruct us sufficiently in the internal dynamics of stars in order to understand the details of the processes responsible for solar and stellar activity cycles.

References

Piddington, J. H.: 1972, *Solar Phys.*, **22**, 3.
Legrand, J. P., and Simon, P. A.: 1991, *Solar Phys.*, **131**, 187.
Snodgrass, H. B., and Wilson, P. R.: 1987 *Nature*, **328**, 696.
Thompson, R. J.: 1988, *IPS Technical Report IPS-TR-88-01*, IPS Radio and Space Services, Sydney, Australia.
Weiss, N. O.: 1993, in *Solar and Stellar Dynamos*, NATO ASI (to appear).

Author index

Subject index

Printed in the United States
By Bookmasters